CHARGE!

CHARGE!

EVERYTHING YOU ALWAYS WANTED TO KNOW ABOUT STATIONARY CHARGERS

William K. Bennett

Golden Alley Press

Golden Alley Press
37 S. Sixth Street
Emmaus, Pennsylvania 18049
www.goldenalleypress.com

This book is set in Minion and Frutiger typestyles
Book design by Terry Guire and Michael Sayre

Printed in the United States of America

CHARGE! / William K. Bennett. -1st ed.
ISBN 978-1-7320276-6-4

Frontispiece photograph © Terry Guire

Photograph of the author © Doreen M. Sutcliffe

Cover design by Michael Sayre

10 9 8 7 6 5 4 3 2 1

CONTENTS

Battery Systems

Battery Chargers

Ripple & Filtering

Output Current Limit

Temperature Effects

Alarms and Other Options

Applications

When Bad Things Happen

Standards & Codes

Appendix A: Factory Testing & Adjustments

Appendix B: Useful Data

Appendix C: Solving Problems

INTRODUCTION

What practical value will you get by reading this book?

You can use it to get a basic understanding of all aspects of secondary batteries[1] and chargers – technologies that play a crucial role in ensuring continuous power availability for critical control systems in electric generating stations and substations, manufacturing, water treatment, and many other applications. You will also obtain answers to technical questions that typically arise for those who specify and purchase dc charging systems.

Why have we emphasized the charging of secondary batteries?

Yes, there are other backup power technologies such as pumped storage or fuel cells that can be used. In many cases, though, these are hampered by cost, complexity, response time, or availability. A storage battery system, when properly designed and maintained, is a simple and reliable technology, with instant availability.

Because of their wide distribution, this manual focuses on charging lead-acid batteries. It discusses several common methods of ac to dc conversion used in battery charging, but especially phase controlled chargers.

What features make it easy for you to find information you are looking for?

Its text is conversational to stimulate interest and make it easy to follow. Questions we would ask, that we think you would want answered, are sprinkled throughout. Start by examining the Table of Contents, as many of the headings are questions.

You do not need to read this manual front to back. Most topics are fully self-contained, and any recommended prerequisites are listed at the start of a topic. If you don't want to know the history of dry cells, for example, skip it.

Finally, icons in the margins draw attention to content that should help you most in understanding and applying the concepts successfully in many practical applications. Glance at the following page for a key to icon meanings.

---

1   In this manual, we frequently use the terms *cell* and *battery* interchangeably. See the definition for battery in the glossary.

KEY TO ICONS

KEY CONCEPT

A central concept to battery and/or charging technology that underpins other subject matter or will greatly assist the reader in using batteries and/or chargers correctly.

TECH TIP

Helps reader select best specification or option for their application, or assists them to get the best performance from their equipment.

SAFE PRACTICE

Emphasizes safe practices, or advises use of manufacturer's safety instructions and/or personal protective equipment (PPE) for an activity.

EXAMPLE

Example problem, calculation or equipment setup with realistic settings and specifications.

ABBREVIATIONS & SYMBOLS

Here is a list of abbreviations used commonly in the text. We have tried to adhere to common standards, such as those defined by the IEEE, NEMA and NIST.

A	Amperes. Also: Aac (ac amperes) and Adc (dc amperes)
ac	Alternating current
Ah	Ampere hours
ANSI	American National Standards Institute
°C	Degrees Celsius (also sometimes called centigrade)
C/(n)	Where (n) is any number, refers to a charge or discharge rate of a cell. In this case, C is the cell's capacity in Ah for the standard discharge time (e.g., eight hours for stationary batteries). As an example, a 100 Ah cell discharged at a current level of 25 A is being discharged at the C/4 rate.
Ca	Calcium, a metal used in lead alloys for manufacturing battery grids
Cd	Cadmium
dB	Decibel; a unit for expressing sound levels (SL) and electrical noise levels
dc	Direct current
EMF	Electromotive force; that is, voltage
EOD	End of Discharge
°F	Degrees Fahrenheit
Fe	Iron. The metal or its oxides is used as cathode material in nickel-iron cells, and in some lithium cells.
ft	Foot, or feet, when modifying a quantity
Hz	Hertz, or cycles per second
IEC	International Electrotechnical Commission
IEEE	Institute of Electrical and Electronics Engineers
in	Inch, or inches, when modifying a quantity
K	Degrees Kelvin, when used to refer to absolute temperature or temperature differences. Temperature differences in Kelvin or Celsius are the same, but the zero point for the Kelvin scale is absolute zero, which is -273 °C.
kA, kV	Represents thousands of amperes or volts. Strictly speaking, they are the unpronounceable kiloamperes and kilovolts. Just say "kay-A" or "kay-V."
kHz	Kilohertz, a thousand Hertz, or a thousand cycles per second
KOH	Potassium hydroxide, an alkaline solution used as an electrolyte in secondary batteries. "K" is the chemical symbol for potassium, from the Latin "Kalium."

kW	Kilowatt, or a thousand watts
Li	Lithium. The metal and its salts are important anode materials for primary and secondary cells.
mA, mV	Thousandths of an ampere or volt; pronounced "milliamperes" or "millivolts."
Mn	Manganese. Oxides of manganese are commonly used cathode materials in primary and secondary cells, such as MnO (manganese oxide) and MnO2 (manganese dioxide).
NEMA	National Electrical Manufacturers' Association
NFPA	National Fire Protection Association, publisher of the NEC, National Electrical Code
Ni	Nickel. NiCd refers to nickel-cadmium cells, and NiMH refers to nickel-metal-hydride cells.
NIST	National Institute for Standards and Technology
Ω	Ohms. You will also see kΩ, thousands of ohms; MΩ, millions of ohms, and mΩ, milliohms.
Pb	Lead, from the Latin, "plumbum"
PWM	Pulse-width modulation
S	Sulfur
Sb, Sn	Antimony, and tin, respectively, both used in lead alloys for manufacturing battery grids (from the Latin names for the metals, "stibium" and "stannum").
SCADA	Supervisory Control and Data Acquisition, a system frequently used by electric utilities and process control industries to monitor and control plant equipment.
SCR	Silicon (or semiconductor) controlled rectifier
UL	Traditionally, Underwriters Laboratories, a NRTL (Nationally Recognized Testing Laboratory); now known simply as UL.
UPS	Uninterruptible Power Supply (or System)
V	Volts. We also write Vac (volts ac) and Vdc (volts dc).
VA	Volt-amperes, or the product of voltage times current. In a dc circuit, this is equivalent to power in watts, but in an ac circuit, VA is usually a larger number than the actual watts. This is explained in more detail in Section 2.4.5.1.
VPC	Volts per cell. In a series string of cells, the VPC is the total string voltage divided by the number of cells.
VRLA	Valve-regulated lead acid, referring to cell or battery construction.
Wh	Watthour
Zn	Zinc

BATTERY SYSTEMS

THIS SECTION GIVES a brief overview of primary and secondary battery systems, their design, construction, and applications. It won't make you an expert, but it will give you a feel for the advantages and disadvantages of each battery type, and how to use them successfully in your application. You'll find that it's helpful to have *Primary Batteries (Non-Rechargeables)* under your belt before tackling *Secondary Batteries (Rechargeables)*.

PRIMARY BATTERIES (NON-RECHARGEABLES)

ALL PRIMARY BATTERIES consist of a positive electrode (the cathode), a negative electrode (the anode), and an electrolyte[1]. The electrolyte is a conductive medium that promotes the transfer of electrons from the cathode to the anode, and then to the external circuit (your flashlight, or whatever). The following paragraphs describe how each battery type is constructed.

ZINC-CARBON CELL	**1.1.1**

History	**1.1.1.1**

In the beginning, all batteries were primary batteries. They're called primary because the energy comes directly from the materials that comprise them, rather than a second source such as a generator (which, of course, gives rise to a secondary battery). The active materials in a primary cell are irreversibly consumed during discharge.

---

1   If you're confused by the statement "The positive electrode is the cathode," go back to the glossary for comprehensive definitions of cathodes and anodes.

The zinc-carbon "dry cell" was developed in the late 1800s, based on research by Georges Leclanché. The name is misleading: carbon isn't part of the cell reaction, and the cell isn't dry.

Construction   1.1.1.2

By 1900, the dry cell (Figure 1a) was made in a zinc can, the anode, with a block of manganese dioxide as the cathode. The electrolyte was an aqueous (that is, water-based) paste of ammonium chloride. Binding the liquid electrolyte in the form of a paste allowed the cell to be used in any position. The manganese dioxide cathode was mixed with carbon powder to improve conductivity, and a carbon or graphite rod acted as the electrical terminal for the cathode (positive terminal). The carbon rod is only a current collector and isn't part of the chemical reaction during discharge.

The development of the dry cell made practical such things as flashlights and portable lighting. The design was a vast improvement over "wet cells," since it had much lower

Figure 1a: A zinc-carbon cell cross-sectional diagram & cross-section of actual cell.

Courtesy of Mcy jerry from https://en.wikipedia.org/wiki/Zinc-carbon_battery under CC BY 2.5 license.

metal cap (+)

carbon rod
(positive electrode)

zinc case
(negative electrode)

manganese(IV) oxide

moist paste of
ammonium chloride
(electrolyte)

metal bottom (−)

self-discharge, and therefore a longer shelf life. The design was further improved by replacing some of the ammonium chloride electrolyte with zinc chloride, which is less corrosive. The "heavy-duty" type of dry cell available today uses primarily zinc chloride in the electrolyte.

Discharge Reaction   1.1.1.3

The chemical reactions of the original Leclanché cell and the "heavy-duty" zinc chloride cell are a little different, due to the presence of zinc chloride in the electrolyte. Both cells, however, consume the zinc anode during discharge, oxidizing it to zinc chloride (and zinc oxide in the heavy-duty cell), and producing water. The Leclanché cell also produces ammonia during discharge. The combination of zinc consumption and water production increases the risk of leakage as a cell discharges.

Applications   1.1.1.4

Dry cells, being the lowest cost primary battery, are ubiquitous, usually being the battery of choice for original equipment in toys, games, and other consumer products. They have a slightly higher initial voltage than other types of zinc-based primary cells.

On the downside, they are more prone to leakage. The zinc can is consumed during discharge, eventually allowing the corrosive electrolyte to leak out and ruin your flashlight. At elevated temperatures, the electrolyte dries out, even if it doesn't leak, reducing long-term capacity.

They're limited to relatively low discharge rates (that is, low current), although they can self-recover from intermittent high-rate discharge, providing a little more life. Just sitting on the shelf, they self-discharge, reducing their service life. Put them in the freezer for longer shelf life but don't forget that they are there. Be sure to warm them up before putting them into service. At low temperatures, capacity is reduced, but the cells usually recover nearly full capacity when they are returned to room temperature.

TECH TIP

Don't believe people who try to sell you a charger for zinc-carbon cells. It doesn't work, and it may be dangerous. All that a charger can do is depolarize any cathode material that has become unavailable to the positive terminal due to an intermittent high-rate discharge; it doesn't "reverse" any of the discharge. Gas is produced internally during attempted charge, which may force electrolyte leakage to increase. Since the zinc can becomes thinner with discharge, a charger increases the likelihood of leakage, with consequent damage to equipment.

The heavy-duty dry cell easily beats the standard version in most discharge tests. Its superiority is in high-rate applications, such as toys; it also has a slight advantage in intermittent duty applications, such as flashlights. But the alkaline-manganese cell surpasses them both.

ALKALINE-MANGANESE CELL   1.1.2

Construction   1.1.2.1

Like the zinc-carbon cell, the alkaline-manganese cell consists of a zinc anode and manganese dioxide cathode. The electrolyte, however, is an aqueous solution of KOH (potassium hydroxide), hence the "alkaline" in the name. Unlike the acidic electrolyte in zinc-carbon cells, the KOH isn't changed in the chemical reaction during discharge. As a result, cell conductivity remains almost constant during normal discharge. Thus, the terminal voltage during discharge remains more constant than for the zinc-carbon cell.

As in the zinc-carbon cell, graphite is mixed with the manganese dioxide cathode to improve conductivity. Also, "sacrificial" zinc powder can be added to the electrolyte to decrease self-discharge; the result is that the shelf life of an alkaline-manganese cell can be up to four years at room temperature, compared to about 1½ years for a zinc-carbon cell.

Applications   1.1.2.2

Typically, an alkaline-manganese cell contains more densely packed active material, giving it greater capacity than a zinc-carbon cell. The lower internal resistance allows a higher rate of discharge. These characteristics make the alkaline-manganese cell more suitable (than zinc-carbon) for high-rate applications such as toys, cameras, lanterns, and locomotives.

While initially more expensive than zinc-carbon cells, the in-service cost of alkaline-manganese cells is typically about half that of zinc-carbon because of higher capacity and longer shelf life.

They're also better than zinc-carbon for applications at low temperatures – which reduce the performance of any battery system. They are superior to zinc-carbon in any application where long standby life is important, such as a flashlight in your glove compartment. In any event, it's important to have a regular schedule for examining and replacing dry cells in portable devices, to prevent electrolyte leakage from damaging your stuff.

TECH TIP

The alkaline-manganese cell is manufactured in a steel jacket. While this makes the cell more resistant to leakage than zinc-carbon, leakage is still possible, especially if the cell is discharged past its useful service point, and/or left in equipment in a discharged condition. The KOH electrolyte that leaks out is hygroscopic (absorbs humidity), and can do serious damage to the equipment by spreading beyond the battery compartment.

Do chargers for primary alkaline batteries work?   1.1.2.3

There are several "chargers" on the market for alkaline-manganese primary cells. Some users of these devices report success in extending the service life of their cells. There are some caveats, though:

- Many users report that cells leak after "charging," usually a few days after they've been returned to the equipment. Reports are inconsistent on which brands are more prone to leakage, or what usage conditions contribute more to leakage.
- Sellers advise you to "recharge" alkaline-manganese cells before they've been discharged more than about 25% to 50%. Kind of frustrates the goal of getting more out of the cell.
- Recharging a primary cell can cause internal gassing, releasing hydrogen. Sellers of these devices give no theory of operation, or any other legitimate technical specs, but it appears that they achieve their magic by charging at a very low rate to minimize gassing. Pressure buildup from gassing and migration of KOH electrolyte to the steel jacket would contribute to the leakage.
- Some sellers claim to be able to recharge primary cells 70 times or more. Actual users report smaller numbers, but even those reports are inconsistent.
- In the absence of any discussion of the chemistry involved, I don't believe that real charging takes place; that is, there is no regeneration of anode material or reversal of the normal discharge chemical reaction.

You may get a little extension of useful time, but you're taking a chance. Do you want to put a recharged cell in your $300 noise-canceling headset, for a flight to LA, and have it leak over Chicago?

LITHIUM PRIMARY BATTERIES   1.1.3

The term "lithium", when describing primary cells, refers to a staggering array of different chemistries. The most common chemistry in use for consumer products is $Li-MnO_2$ (lithium-manganese dioxide), used in such popular button or coin cells as CR2016 or CR2032. You probably have one of those in your watch or your car's keyless entrance transmitter.

Lithium is an attractive anode material because of its light weight, yielding cells with high energy and power densities. The $Li-MnO_2$ cell has an open-circuit voltage of 3.3 V, allowing it to replace two zinc-based cells in series. Other chemistries, such as lithium-iron or lithium-iron sulfide, can be used to design a cell with 1.5 V open circuit, to be a drop-in replacement for zinc-based cells. Most of these chemistries use metallic lithium as the anode.

Lithium, of course, has one disadvantage: on exposure to air, it will combust violently, usually described as an explosion. It also reacts violently with water, so lithium primary cells must use a non-aqueous electrolyte. The electrolyte has a high internal impedance, limiting the usefulness of most lithium primary cells to low-drain applications.

"Exploding" laptop computer and mobile phone batteries (actually, they only smoke and burn, destroying the device) have had extensive news coverage in recent years. While these are rechargeable batteries, not primary, they have given lithium some bad publicity. Still, this has not prevented lithium from becoming the most popular choice for portable devices. Though some primary lithium chemistries may explode spontaneously if they are accidentally short-circuited, they are not generally used in consumer products.

Just to be safe, never try to open a lithium cell.

SAFE PRACTICE

Applications (Li-MnO2) 1.1.3.1

This lithium primary cell is characterized by very high energy density and low self-discharge rate at normal room temperatures. It's good for low-rate applications where long service and/or standby life are required, such as watches. It's also capable of intermittent high-rate discharges. It's proven to be very reliable; just don't take a hacksaw to it.

Applications (Li-FeS2) 1.1.3.2

Lithium-iron disulfide cells are suitable for high-rate applications, and can be used to replace alkaline-manganese cells, since their terminal voltage is about 1.5 V. They have about three times the energy and power densities of alkaline-manganese, a flat discharge curve, low internal resistance, and very low self-discharge. Shelf life at room temperature may be measured in decades. They're ideal for both continuous and intermittent high-rate applications, such as cameras and toys. Of course, they also cost about three times as much as alkaline manganese. That seems to limit them to applications where they excel over other types: long standby life, followed by a high-rate discharge.

OTHER PRIMARY BATTERIES   1.1.4

There are several other primary battery systems in common use. These generally have a small number of specific applications.

Zinc-Air   1.1.4.1

As the name implies, zinc is the anode, and the cathode uses oxygen for the chemical reaction. Since zinc is a relatively light metal, the cell has a high energy density. Because zinc is less reactive in air than lithium, zinc-air is a much safer cell than lithium. Small Zn-Air cells are widely used for hearing aids and cameras.

Silver Oxide   1.1.4.2

They're commonly called "silver cells," but, of course, zinc is the anode. Did you expect anything different? The electrolyte is potassium hydroxide or sodium hydroxide. Although they have high energy density, their use is limited to small button cells (calculators, watches, and the like) because of the high cost of silver, or to large batteries used by the military, where cost may not be an obstacle. Some silver oxide cells can be recharged.

Mercury Cells   1.1.4.3

The mercury cell is based on a mercury oxide cathode (or a combination with manganese dioxide), and – guess what? – a zinc anode (by now, you've bought some zinc mining stock, right?). The electrolyte is potassium or sodium hydroxide. The mercury cell has the advantage of constant output voltage over its service life, and a long (10-year) shelf life. It was used in, for example, cardiac pacemakers, cameras, and hearing aids.

It's academic, though, since they are no longer commercially available because of the mercury content. They're described here to complete the historical record.

COMPARING PRIMARY CELLS   1.1.5

Table 1a shows a comparison of features for commonly available primary cell types.

Cell Type	Nominal Voltage	Electrolyte	Capacity (Zn-Carbon=1)	Discharge Slope	Self-Discharge	Low Temp Performance	Peak Current
Zinc-carbon	1.5	Ammonium chloride + zinc chloride	1.0	High	Moderate	Poor	Moderate
Heavy-duty zinc-carbon	1.5	Zinc chloride, primarily	1.25	Moderate	Low to moderate	Poor	Moderate to high
Alkaline-manganese	1.25	KOH (potassium hydroxide)	5	Moderate	Low	Fair	Moderate to high
Li-MnO$_2$ (or coin cell)	3.0	Organic	*	Moderate	Low	Good	Low
Li-FeS$_2$	1.5	Organic	10-15	Flat	Very low	Very good	High

* There is no direct comparison between Li-MnO$_2$ and alkaline button cells. There aren't any crossover sizes, and the lithium cell has twice the voltage of the alkaline. We would expect the lithium cell to have about twice the mAh capacity of alkaline for the same volume, which means it has about 4 times the energy density by volume.

SECONDARY BATTERIES (RECHARGEABLES)

SECONDARY BATTERIES ARE so named because the original source of the energy they provide is from a second source, such as an ac-powered rectifier, rather than the materials that comprise the battery, as in a primary cell. A much more descriptive name is rechargeable battery. All secondary batteries are rechargeable, while most primary batteries are not rechargeable.

ALKALINE BATTERIES   1.2.1

Alkaline batteries are those that use an alkaline electrolyte, usually KOH (potassium hydroxide). Virtually all commercially available alkaline rechargeable batteries are based on compounds of nickel. The names of the various chemicals used in battery design can be arcane, so bear with us.

Nickel-Cadmium Batteries   1.2.1.1

The nickel-cadmium (or NiCd) chemistry has been around since the late 1800s (when it was developed in Sweden), and accounts for most of the small rechargeable cells sold in the United States, amounting to several hundred million cells each year. Most of these, in turn, are sealed cylindrical cells in the AAA through D sizes with nominal terminal voltage of 1.25 V. Larger flooded (unsealed) cells are used primarily in Europe and other areas, while lead-acid batteries are preferred in the U.S. Also see *Large NiCd Batteries* below.

In the small sealed cell design, the electrode layers, along with separator layers, are spirally wound (like a jelly roll) within a cylindrical shell. Cadmium metal (Cd) forms the anode (negative terminal) in a fully-charged NiCd cell, with nickel-oxide-hydroxide (NiO(OH)) forming the cathode. The electrolyte is a solution of potassium hydroxide.

NiCd cells have excellent high-rate discharge capability and can be discharged to zero volts without permanent damage (although cell reversal should be avoided). They have high power density (about twice that of lead-acid batteries), high cycle life, and good performance at low temperatures. It is this last feature that is valuable in stationary and military applications.

The major disadvantage is the presence of cadmium, a toxic metal. For this reason, NiCd batteries are being restricted in the European Union. Other negative features are a higher cost than lead-acid and a possible "memory" effect (at least for small cells), which may reduce usable capacity in some applications. NiCd batteries have a moderate self-discharge rate.

Discharge Precautions   1.2.1.1.1

A single NiCd cell can be discharged to zero volts without damage, and in fact can be stored in a completely discharged state. Since the KOH electrolyte isn't changed by the discharge reaction, the specific gravity doesn't change. The freezing temperature is -25 °C or lower; most large industrial batteries are rated for -40 °C operation.

Discharging a string of NiCd cells that are in series, however, carries the risk of cell reversal (SECTION 1.4.7). Manufacturing tolerances inevitably result in some cells having lower ultimate capacity than others. Those with the lowest capacity in a series string will be discharged to zero first; as the string continues to discharge, those cells will be reversed. Most small NiCd cells will tolerate a reversal of up to a few tenths of a volt, but beyond that, internal gassing may cause permanent damage. To be safe, a series string of cells shouldn't be discharged below about 0.9 VPC (volts per cell), and certainly not below 0.5 VPC.

Charge Precautions   1.2.1.1.2

Large NiCd cells can safely be charged at the C/4 rate, where C is the capacity in ampere hours. For example, a 100 Ah battery can be charged at 25 amperes, provided that the terminal voltage is carefully monitored. The charge should be terminated when the gassing potential is reached. The cell temperature should also be monitored, since the gassing voltage decreases as the cell temperature increases.

Small sealed cells should be charged at no more than the C/10 or C/20 rate. As the cell approaches full charge, gas is generated internally. The cell is designed so that the negative electrode has a slightly higher capacity than the positive, and this allows the cell to absorb the gas at controlled charging rates. However, internal pressure will increase if the charging rate is excessive. With a high internal pressure, the cell will vent to release the gas. If the vent doesn't reclose, the electrolyte is exposed to the atmosphere, and the cell eventually dries out, losing capacity. A non-reclosing vent can also leak electrolyte, damaging the host equipment. Repeated overcharge, even with a reclosing vent, eventually reduces capacity and cycle life.

So-called fast chargers for NiCd cells usually have a provision for monitoring cell temperature. NiCd cells have a negative temperature coefficient, so that as the cell heats up toward the end of the charge, the terminal voltage decreases. This increases the charge current, causing more heating, and a vicious cycle is generated ("thermal runaway"). A fast charger that monitors the cell temperature can terminate charging when a critical temperature is reached. A "smart" fast charger can also detect a decrease in on-charge voltage, and terminate the charging to prevent thermal runaway.

Large NiCd Batteries   1.2.1.2

NiCd batteries aren't limited to portable applications. Large NiCd batteries, up to a thousand ampere hours or more in capacity, are used in industrial applications. In most designs, the electrode materials are contained within steel pockets (the "pocket plate" design), rather than being spirally wound as in small cylindrical cells. Most of these are flooded cell designs, which experience water loss during long-term float charging in stationary applications, due to electrolysis (SECTION 1.5.2.3). To reduce maintenance requirements, some versions are manufactured with pressure vents and excess negative electrode capacity, which help to reduce water loss by recombining the evolved hydrogen and oxygen.

NiCd batteries are extremely rugged, both mechanically and electrically, with long cycle life and long shelf life. They can be stored in a discharged condition. The downsides, of course, are cost and the inclusion of cadmium.

The presence of cadmium classifies NiCd batteries as hazardous waste. All NiCd batteries should be recycled.

SAFE PRACTICE

Nickel-Metal-Hydride Batteries   1.2.1.3

Nickel-Metal-Hydride (NiMH) cells are the new kid on the block, relatively speaking. Their development was spurred in the 1970s for automotive applications because of their high volumetric and weight energy density, a significant advantage over NiCd. They obviously have the added advantage of eliminating the cadmium.

Characteristics   1.2.1.3.1

Small sealed cells based on NiMH technology are available in AAA and AA sizes. C and D sizes are available, but consumer grade cells are usually equivalent to AA cell capacity, just in larger jackets. The NiMH cells have two times, or greater, the capacity of equivalent NiCd cells. Their nominal terminal voltage is 1.25 V, similar to NiCd, so they can be used as drop-in replacements for NiCd cells. A variety of electric vehicles (EVs), including hybrids, use larger NiMH batteries, up to several Ah, though these are beginning to be outnumbered by lithium-ion batteries.

Discharge Profile   1.2.1.3.2

The nominal terminal voltage on discharge is 1.15 to 1.25 V (depending on discharge rate; the discharge voltage should be about 1.25 V at the C/5 rate, and lower at higher rates). This voltage doesn't change significantly with temperature, although capacity is reduced at low temperatures (to as little as 20% at -20 °C); by comparison, NiCd cells perform better at temperatures down to -20 °C. At elevated temperatures (up to 50 °C), the discharge voltage and capacity remain nearly constant.

The terminal voltage holds up well until the end of discharge, when it collapses quickly, to about 0.8 to 1.0 V. Since the discharge voltage is constant to this point, the voltage can't be used to indicate the state of charge. Discharging a single cell to zero volts won't cause permanent damage to a NiMH cell, but deep discharges of series strings should be avoided.

Traditionally, NiMH cells have had a high self-discharge rate, losing up to 50% of their charge in a few months. This performance has been improving as the technology has matured. This effect increases at higher temperatures; it's a good idea to store charged NiMH cells in the refrigerator if initial capacity is important. Return them to room temperature before putting them in service. In a normal cycling application, the capacity lost through self-discharge is recovered on the first (or maybe the second) recharge.

Because of this self-discharge, NiMH cells are recommended primarily for high discharge rate applications, such as digital cameras. In low drain equipment, such as clocks, service life and shelf life are similar.

In low drain applications, or where long shelf life is required, a better choice might be a primary alkaline-manganese cell, or, for low temperature applications, primary lithium-iron disulfide. NiCd cells could be used if the ability to recharge is important, because of NiCd's lower self-discharge rate compared to NiMH.

The "memory effect" that may plague small NiCd cells is absent in NiMH cells, so repeated shallow discharges won't affect their performance in standby or shallow cycling applications. They aren't particularly suited for standby applications, though, because their self-discharge rate is higher than for NiCd.

There is a natural variation in the self-discharge rate between cells. Because of this, and the higher self-discharge rate, there is a greater risk of cell reversal when series connections of cells are discharged. As in the NiCd system, even slight cell reversal can cause permanent damage if repeated often. For consumer products, the best practice is to stop the discharge as soon as the battery performance is noticeably degraded.

Life expectancy is no more than about five years due to degradation of the active materials and separators. Cycle life can be 500 discharge-charge cycles or more, provided that proper charging constraints are observed. That's next.

Charging the NiMH Cell or Battery   1.2.1.3.3

You should not use your old NiCd charger to charge a NiMH cell, unless you are sure that it will charge at the C/40 rate or lower. The best way to charge a NiMH cell is with a "smart" charger, which monitors the cell voltage and temperature to terminate charging when the cell is fully charged. Overcharge can cause severe damage. If you buy a kit of NiMH cells, along with a charger, the charger is designed specifically to recharge the NiMH cells.

Available smart chargers for NiMH cells can recharge a cell completely in two or three hours, far faster than the 12–14 hours required for NiCd. Never use a constant voltage charger. If you use a trickle charger at the C/20 rate or higher, it should be controlled by a timer to terminate the charge after the appropriate time. See *Under the Hood* below for an explanation of charging requirements.

Under the Hood   1.2.1.3.4

As in the NiCd cell, the cathode (positive electrode) is nickel oxide hydroxide (sometimes referred to as nickel hydrate). The anode is a compound of rare earth metals or transition metals with nickel, cobalt or other metals, using multiple recipes. The metals are reduced to hydrides in the charged state. In this chemistry, a "hydride" is considered to be an alloy of the metal with hydrogen, rather than a molecular compound. The metal recipes are chosen for their unique ability to adsorb huge quantities of hydrogen – many times their own volume.

Like small NiCd cells, small NiMH cells are constructed with excess capacity at the negative electrode. As the cell approaches full charge, the cathode (positive) becomes fully charged first, and starts to generate oxygen. The oxygen migrates to the negative electrode, where it combines with hydrogen to form water. This recombination prevents a dangerous rise in internal pressure. The recombination works, however, only at very low charging rates (C/40 or lower). For continuous trickle charge, some manufacturers recommend rates as low as C/300. For a cell rated 1500 mAh, this translates to a charge current of 5 mA. To avoid a ridiculously long charge time, it makes sense to invest in a smart charger. Read on.

At higher rates, another method of charge termination is required. The end-of-charge voltage characteristic of a NiMH cell has a negative slope, meaning that the on-charge voltage will decrease slightly when the cell becomes fully charged. In addition, internal cell pressure and temperature will rise markedly. A smart NiMH charger monitors the cell voltage and terminates the charge when it detects the negative end-of-charge slope. Some smart chargers also monitor cell temperature, with a temperature sensor built into the battery clip contacts.

It's important to limit the internal pressure during recharge. If the pressure rises high enough to vent the cell, the cell loses water from the electrolyte and capacity is reduced. There is also the risk that electrolyte (usually KOH) released through the vent can damage host equipment.

SAFE PRACTICE

Recycling   1.2.1.3.5

A NiMH cell lacks cadmium, so it's tempting to dispose of it in normal household (or industrial) waste. If possible, NiMH cells should be recycled, to recover the metallic components. Recycling also disposes of the KOH electrolyte properly, since in large quantities it can cause environmental damage. Large NiMH batteries, such as from EVs or hybrids, should be recycled through vehicle dealers or the original battery suppliers.

SAFE PRACTICE

LEAD-ACID BATTERIES   1.2.2

Lead. Chemical symbol Pb (from the Latin plumbum). Atomic number 82, atomic weight 207 grams/mol. Lead is an abundant heavy metal, easily and cheaply extracted from ore. It was used by the ancient Egyptians and was widely used for plumbing in the Roman Empire, leading to one of the theories regarding the empire's collapse.

Lead is also poisonous. It accumulates in soft tissues and bones and causes blood disorders and brain damage. Therefore, take all spent lead-acid batteries to a disposal or recycling facility that takes measures to prevent the lead from becoming a future health hazard.

SAFE PRACTICE

An electrical couple made from lead (as a negative electrode) and lead peroxide (as a positive electrode), using dilute sulfuric acid as an electrolyte, forms a useful secondary battery; in fact, it was the first commercially-developed secondary battery in modern history, invented by Gaston Planté in 1859 (for an interesting ancient possibility, search the internet for Babylonian batteries). The simplified charge and discharge reactions are:

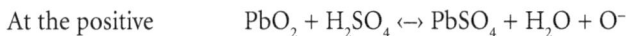

At the positive $PbO_2 + H_2SO_4 \longleftrightarrow PbSO_4 + H_2O + O^-$

At the negative $PB + H_2SO_4 \longleftrightarrow PbSO_4 + 2H^+$

The double-headed arrows indicate that the reactions are reversible. Discharging reads left to right, and charging reads right to left. The ions O^- and H^+ combine to produce a water molecule. Note that during discharge, both the negative and positive electrodes are converted to lead sulfate. The consequence of this reaction is that the sulfuric acid electrolyte, H_2SO_4, is "consumed" by discharging, changing into water. There are two undesirable results. The first is that the cell terminal voltage decreases, since the cell voltage is proportional to the concentration of acid. Second, the electrical resistance of the electrolyte increases (acid is a pretty good conductor, but water is a poor conductor), decreasing the available terminal voltage of the cell when load current is flowing. The terminal voltage of the cell drops off slowly at the start of the discharge, but more rapidly toward the end of discharge.

For a deeper look into the reactions, try searching the internet for "lead-acid battery." Wikipedia and several other internet sites have more information.

The lead-acid battery has miserable energy density by weight, poor energy density by volume, and pretty good power density, due to its ability to deliver high discharge currents. Since the electrolyte is consumed during discharge, the cell has poor voltage regulation (a steep voltage discharge curve). The cell, with a nominal open-circuit voltage of 2.0 V, can't be discharged below about 0.9 V without risking permanent damage, which would be evidenced by reduced capacity. It has terrible performance at low temperatures and may freeze if fully discharged. Most designs have poor cycle life. Overcharging can release toxic gases, and cause irreversible corrosion of the plates, again reducing capacity. Continuous undercharging can cause irreversible sulfation or hydration, resulting in physical damage to the cell plates. Lead-acid cells have a self-discharge rate of up to 10% per month. Because of the toxicity of lead, battery disposal requires special handling. Recycling is mandatory in most states. Despite all this, *the lead-acid battery accounts for 90% of all rechargeable batteries, worldwide,* not counting small cells for portable equipment, which are usually NiCd, NiMH or lithium-ion. It's tough to beat the fact that it's abundant and … cheaply extracted from ore.

Automotive SLI (starting, lighting, ignition) batteries make up about half of all rechargeable batteries sold. Most are lead-acid. Other major applications for lead-acid include traction batteries (fork lift trucks, golf carts, etc.), telecommunications, switchgear and process controls, and uninterruptible power systems. Smaller batteries are found in emergency lighting and alarms, small vehicles such as scooters and bicycles, and my battery-operated lawn mower.

Because of their poor energy to weight ratio, lead-acid batteries are not likely to find widespread application in electric passenger vehicles. Individuals who build their own electric vehicles generally use lead-acid because of cost and availability. The Chevrolet EV1 initially used a lead-acid battery, but that was replaced by a NiMH battery during manufacture of the second generation of the vehicle.

Remaining sections of this publication are concerned primarily with industrial lead-acid batteries, chiefly stationary batteries.

Construction   1.2.2.1

Pasted-Plate Cell   1.2.2.1.1

A pasted-plate lead-acid cell (Figure 1b) is made from flat plates of electrode material,

Figure 1b: Construction of pasted-plate cell. Courtesy Tony Ortiz, HindlePower, Inc.

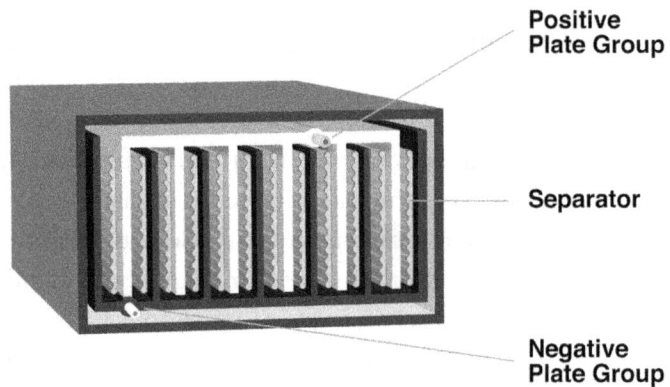

Positive Plate Group

Separator

Negative Plate Group

with the positive and negative plates stacked alternately in a "jar," which may be made of glass or plastic. There is always one more negative plate than the number of positive plates; thus, there is always an odd number of plates in a cell. The negative plates are on the outside of the stack of plates at both ends of the jar. The alternating positive and negative plates in a stack are separated by a non-conductive material, such as glass fiber or porous plastic.

A plate consists of a grid cast from molten lead, usually alloyed with another metal to increase its strength. The grid is designed with pockets into which a paste of active material is pressed, resulting in a "pasted-plate" cell. The lead grid isn't ordinarily involved in charging or discharging the cell; it's only there to conduct the current in or out and contain the active material.

Various alloys have been developed for manufacturing lead grids. The two most common are antimony (chemical symbol Sb) and calcium (Ca). Lead-antimony grids may also have some selenium in the alloy, which helps to reduce the amount of antimony needed. These cells are frequently marketed as "selenium cells," and are sometimes called "low-antimony" cells. See the boxout *Antimony or Calcium?* for more information on their characteristics.

ANTIMONY OR CALCIUM?

Lead is a soft and relatively weak metal and is difficult to cast into grids in pure form. A pure lead grid isn't strong enough to maintain its geometry in a cell assembly.

Antimony was one of the first metals to be alloyed with lead to form a strong castable alloy. It produces a battery capable of frequent cycling, but because of low gassing voltage, it requires more water addition when used in float service. During float charging, antimony is lost from the grid alloy and expelled as gaseous stibine (antimony hydride), or migrates to the negative plate, causing local discharging. This increases water requirements, and eventually causes grid or plate failure.

Lead-calcium alloys were developed to overcome some of the shortcomings of the antimony alloy, especially high water usage. Because it has higher gassing potential, calcium alloy cells require far less maintenance in float service, but do not cycle well. Calcium alloy grids are used in VRLA batteries and some float service batteries, such as stationary applications. During long-term float service, the calcium is subject to corrosion, which leads to plate failure.

Alloys with low antimony, sometimes with the addition of selenium, overcome the gassing problems of high-antimony alloys, and keep the cycling ability of the antimony alloy. Another alloy is lead-tin, which shows promise for reducing thermal runaway in VRLA batteries.

Internally, each set of plates (positive and negative) is interconnected with lead buses. Each set is then connected to a lead or lead alloy post at the surface of the jar to form the external cell terminal. The jar is filled with a dilute solution of sulfuric acid.

Tubular Construction   1.2.2.1.2

A tubular battery uses a construction different from a pasted-plate battery for the positive plate. The lead grid, instead of having pockets to hold the active material paste, has a row of vertical spines, which are in turn covered with a tube of (usually) glass fiber material. The tube is filled with active material in powder form. The spines make intimate contact with the active material, conducting the current in the same manner as the lead grid in the pasted-plate cell. Because the active material is captive in tubes, rather than being on the surface of flat plates, it's more resistant to shedding, and the positive plate has a potentially longer life, capable of more charge-discharge cycles. But because the volumetric density is lower, peak current capability isn't as high as the pasted-plate construction. Manufacturing cost is also higher.

Flooded & Sealed Batteries   1.2.2.2

Traditionally, all lead-acid batteries were "flooded," (Figure 1c) that is, the plates were immersed in liquid electrolyte, with excess electrolyte above the tops of the plates. During its lifetime, the cell loses electrolyte through evaporation, electrolysis and grid corrosion. The

Figure 1c: Example of flooded cells, showing vent caps used to refill electrolyte losses, & fill level markings

Vent Cap

Safe Fill Level Region

space in the cell above the plates, an electrolyte reservoir, allows the cell to be replenished with water to restore the optimum amount of electrolyte. This obviously requires an access port at the top of the cell, which is normally closed with a vent cap. The vent allows the escape of gas generated during charging. Another name for a flooded cell is vented cell (see the definitions in the Glossary).

So, if there is a vented cell, there must be an unvented cell, right? Well, yes, relatively speaking. While we haven't covered charging yet (although you're welcome to read ahead if you want), you should know that fully charging a lead-acid battery always results in gassing: evolving hydrogen and oxygen by electrolysis. Always. During charging, hydrogen is evolved at the negative plate, and oxygen at the positive.

Since we're breaking up water, H_2O, into its constituent gases, hydrogen and oxygen, they'll be evolved in a stoichiometric ratio. This is about the same as saying "explosive."

Whether the battery is vented or not, we must make provision for the evolved gases. In a flooded cell, the vent caps usually have flame arrestors. If the gas on the outside of the battery becomes concentrated enough, and is accidentally ignited, the flame arrestor prevents the explosion from propagating to the inside of the cell, and vice-versa.

Now about that unvented cell. The flooded battery requires maintenance, of course. Every so often, you have to "top up" the electrolyte level. A person, at the site, in the battery room. With a big container of water. Someone must clean the tops of the jars, tighten the connections, and if your policy is really ambitious, check the voltage or specific gravity of each cell. With more sophisticated equipment, this person could also check the impedance of each cell, and/or the intercell connection resistances.

Many users, perhaps including you, would like to reduce that maintenance requirement, especially the part about adding water to the cells. It may take two or three minutes to top off a cell manually, if you do it right. It takes only a few seconds to check the connectors on each cell (be sure to use an insulated wrench).[2] For you, manufacturers developed "sealed" or "maintenance-free" batteries. Of course, they are neither sealed nor maintenance-free, but they're a step in that direction. They aren't truly unvented; the vents are internal, where you normally can't get at them. In most types, you can't add water, even if you want to. The correct term is valve-regulated; we'll use the industry standard abbreviation VRLA, for valve-regulated lead-acid.

KEY CONCEPT

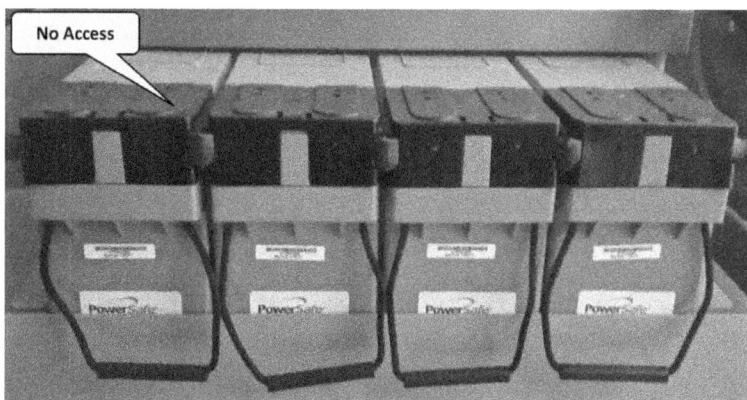

Figure 1d: Example of valve-regulated lead-acid (VRLA) batteries, often called "maintenance-free" batteries

SAFE PRACTICE

2   While we're on the subject, you also need to give careful attention to safety when installing or maintaining batteries. Use safety goggles and other protective equipment. Have handy the necessary materials to clean up spills – sodium bicarbonate or ammonia for lead-acid, and boric acid for alkaline batteries. The battery manufacturer can provide extensive information on safe installation and maintenance.

True: VRLA batteries (Figure 1d) are not flooded. They have no external vent caps. They have no electrolyte reservoir. They can be operated on their sides, as well as the normal right-side-up position; some can be operated upside down. Most are classified by the Department of Transportation as "nonspillable," and are cleared for air transport.

False: VRLA batteries never gas.

Of course, VRLA batteries produce gas during charging. What we need is a means to cope with the evolved gases without letting them escape from the battery. We accomplish this with a vent system designed to seal the battery at atmospheric pressure and slightly above. The cell is constructed so that, at slightly elevated pressures, the evolved gases migrate to the opposite electrode, where by magic they are recombined into water (see the boxout *Recombination Magic*). If something goes wrong (too much charging current, for example), the cell can vent to release excess pressure, but this inevitably leads to electrolyte loss and reduced capacity. So, to qualify the previous statement that we marked as false, we can say, then, that "VRLA batteries won't release gas, when operated and charged properly in a controlled environment."

RECOMBINATION MAGIC

Oxygen at the positive plate migrates to the negative plate, where it reacts with the (sponge lead) active material to produce PbO (lead oxide). In turn, PbO reacts with the electrolyte to make $PbSO_4$ and H_2O, restoring the water lost through electrolysis. Float current recharges the $PbSO_4$ back to the sponge lead active material at the negative plate.

Hydrogen evolution at the negative plate is suppressed by the action of the oxygen migrating from the positive. Any excess hydrogen is recombined by a catalyst in the cell. At carefully controlled float charging rates, recombination efficiency is nearly 100%.

In both gel cells and AGM cells, microchannels or pores in the gel or glass fiber separator facilitate the migration of oxygen to the negative plate.

"Gel Cells"   1.2.2.2.1

Gel/Cell (note the forward slash) was originally a trade mark of Globe Union. Today, the generic term gel cell is almost universally used to describe a sealed battery with a gelled electrolyte. Since the electrolyte isn't a liquid, the cell is very resistant to leakage, and can be operated in any position (although some manufacturers recommend against upside-down operation). Gel cells are very rugged, and capable of moderately high peak discharge currents. The grid alloy is lead-calcium. Because the electrolyte quantity is limited (starved electrolyte) and has less mobility than the free electrolyte in a flooded cell, the cell has some inherent protection against damage from high-rate deep discharges.

AGM (Absorbed Glass Mat) Cells   1.2.2.2.2

Like the gel cell, the AGM cell is virtually leakproof and can be operated in any position. The electrolyte is a liquid absorbed in a glass fiber separator; the plates and separators form a very tight package. This gives the AGM cell the capability for high peak discharge currents and leads to a lower Peukert constant than a gel cell[3] . Thus, the capacity of an AGM cell decreases less during high rate discharges than does the gel cell. The AGM is also a better performer at low temperatures.

Perfect, right? Well, not exactly. VRLA cells are less tolerant of elevated temperatures than flooded cells. At higher operating temperatures, a temperature-compensated charger is highly recommended (I'd like to say mandatory). Even with temperature compensation, a VRLA cell's float life will be decreased by about half for each 10 °C above room temperature (a factor that applies to all lead-acid batteries).

» *Q: Does that reduction in life apply to nickel-cadmium batteries, too?*

 A: Yes, but not to the same extent, according to NiCd battery manufacturers. For each 10 °C above room temperature, expect to lose about 20% of the battery's rated lifetime.

Cylindrical Cells   1.2.2.2.3

Cylindrical lead-acid cells were developed around the 1970s to overcome some of the problems with high-rate discharge at low temperatures. The plates are extremely thin lead-tin alloy, pasted with active material, and then spirally wound with an AGM separator. The sealed, high-pressure vent construction is like small NiCd and NiMH cells. They're available in standard D sizes and larger, with capacities from a few Ah to a few tens of Ah. They can be used in portable equipment because of their resistance to leakage but are heavier than NiCd or NiMH cells with the same capacity. Lithium batteries will eventually replace most heavy metal batteries in portable applications.

---

3   Peukert studied the relationship between cell capacity and discharge rate. We cover that subject in *Discharging Stationary Batteries*, in Section 1.4.

Planté Cells	1.2.2.2.4

The Planté cell design is an attempt to get back to basics – using pure lead as the active material in the positive electrode. A pure lead grid is supported by a structure of lead alloy (usually lead-antimony). The grid is grooved, scored or etched to increase the surface area, or may be cast with openings which are filled with spiral pure lead "rosettes." The lead plate and/or the lead rosettes are oxidized to lead peroxide during the forming process to produce the positive active material. Plates using the rosette design may be referred to as a Manchester design, or a Manchex™ design (Exide/Hawker trade name).

In a Planté cell of either design, the negative plate is usually the standard pasted-plate design.

The Planté design has very long service life, low self-discharge rate, and relatively low energy density. Because it also has a higher manufacturing cost, it's best in applications where long unattended float service life is required. Although the Planté design performs well in cycling applications, more maintenance would be required, due to the need for occasional equalization charging (for more on this subject, see Section 1.5.5).

LITHIUM SECONDARY BATTERIES   1.2.3

We refer to lithium-ion batteries, rather than pure lithium, when discussing rechargeable lithium systems. Remember that in primary systems, lithium metal is usually used for the anode, and since the lithium is "consumed" during discharge, the reaction is non-reversible. To develop a rechargeable lithium cell, the lithium metal was replaced by an alloy or compound of lithium; as is true for primary cells, there are many competing chemistries. The anode is commonly carbon or a lithium alloy; common cathode materials are lithium cobalt oxide, lithium manganese oxide, and lithium nickel oxide. Lithium ions are "intercalated" in the structure of the anode and migrate to the cathode during discharge. Charging reverses the migration.

Lithium is the lightest solid metal, and the most electrochemically active. It's an abundant metal, but more expensive to extract from ore than lead because the processes must be non-aqueous and the metal can't be exposed to air. Lithium ion cells have the highest energy densities, easily four times those of lead-acid or NiCd. Lithium polymer cells, also called lithium ion polymer, have the electrolyte solidified in a polymer; these have extremely high power densities, as well as very high energy densities. Lithium cells also have high open circuit and operating voltages, between 3.5 and 4.0 V for most designs.

The combination of abundant raw materials, light weight, and high energy density makes the lithium secondary battery the obvious choice for the battery of the future. In 2010, the U.S. government began stimulating manufacturing start-up for several companies in the industry with the intent of lowering manufacturing costs and expanding its business potential. Lithium ion's initial costs to produce in 2016 were still above $500/kWh – more than 4 times the cost of a standard lead-acid flooded battery, and more than 1.5 times the cost of a premium, maintenance-free VRLA gel cell. Now that large scale manufacturing facilities are ramping up production, especially in China, lithium cell cost may decrease to the point where it surpasses lead-acid, not only for applications that require low weight and mobility such as for cellphones, but also for stationary applications discussed in this book.

Lithium batteries are normally not sold as individual cells, but are integrated by the manufacturer in an assembly with protective circuitry, to ensure that the battery always operates within safe voltage and current limits.

Discharge Profile   1.2.3.1.1

The discharge voltage curve has a moderate slope – greater than NiCd, but better than lead-acid. The discharge is normally terminated at 2.5 VPC (volts per cell). Lithium batteries must not be discharged below about 2.0 VPC, or permanent damage will result. The discharge cutoff is usually part of the control circuitry in the battery assembly, which is part of each manufacturer's battery design. Controlling charge termination also avoids cell reversal.

The discharge voltage is moderately dependent on discharge rate and on temperature (higher with higher temperature), but even at 0 °C, the cell voltage is usable. Lithium ion batteries will operate at fairly high temperatures, but most designs are limited to 100 °C. Don't try that with lead-acid.

Charging Lithium-Ion Batteries   1.2.3.1.2

Recharging the lithium-ion battery requires careful control to keep the battery in the safe operating area. Charging is usually performed with constant current until a critical cell voltage is reached. Charging then continues at that voltage until the charge current drops below a threshold value, typically 3% of the initial charge current. Compared to NiCd or NiMH batteries, lithium-ion batteries have a relatively low self-discharge rate, but an occasional "topping off" charge may be required.

Lithium-ion batteries have no memory effect. Depending on the chemistry used, cycle life can be several hundred to a thousand or more discharge-charge cycles. As the battery ages, the internal impedance rises, and eventually the discharge terminal voltage becomes unusable.

Recycling   1.2.3.1.3

All lithium batteries should be recycled if possible. While lithium isn't a hazardous waste, many lithium battery types contain some heavy metals (such as cobalt), or other hazardous materials.

OTHER SECONDARY BATTERIES   1.2.4

Nickel-Iron Batteries   1.2.4.1

The NiFe (nickel-iron) battery was commercialized primarily for powering electric vehicles in the early 20th century by Thomas Edison; it became known as the "Edison Battery" although it was invented by Ernst Waldemar Jungner who also brought you the NiCd battery. The anode (negative electrode) is a pocket grid of iron, the cathode is nickel hydroxide, as in the NiCd battery, and the electrolyte is potassium hydroxide.

The major disadvantages to the NiFe battery are low volumetric energy density, low power density, low (and slow) charge efficiency, and high self-discharge rate. It has a high gassing rate during recharging, and therefore requires more maintenance than other secondary batteries. But its great advantages are an extremely long cycle life, high tolerance to abuse, and good high temperature performance. Expected service life is at least 20 years. The anode and cathode active materials are less toxic than those in NiCd and lead-acid batteries.

The NiFe battery is used today in mining and some solar power applications, where its high cycle life is an advantage.

Silver-Cadmium Batteries   1.2.4.2

The silver-cadmium battery substitutes silver oxide for the nickel hydroxide cathode material used in a NiCd battery. Compared to NiCd, the silver-cadmium battery is characterized by higher energy density, higher power density, lower self-discharge rate, equivalent cycle life, and better low- and high-temperature performance.

SAFE PRACTICE

Perfect, right? Not when you consider the cost of silver. Also, they still have environmental problems: silver and cadmium are regulated hazardous waste, according to the EPA. This requires recycling or expensive disposal at the end of life. Batteries containing silver are normally recycled anyway because of cost.

They're still used in a few NASA and military applications, but they will probably be phased out because of the cadmium content, in favor of silver-zinc, NiMH, or lithium technologies.

Silver-Zinc Batteries   1.2.4.3

Silver-zinc rechargeable batteries are also primarily in use by the military, again because of cost barriers. The silver-zinc battery comprises an anode of zinc, a cathode of silver oxide, and potassium hydroxide as the electrolyte. They have very high energy density compared to nickel technologies, and have higher energy density than lithium ion batteries. But their cycle life is limited by anode dendrite growth, which requires special techniques to control, including the addition of mercury.

Despite their cost, at least one manufacturer has commercialized a silver-zinc secondary battery for hearing aid applications. The manufacturer had tried to develop a larger secondary battery for portable computers, but that effort has apparently been abandoned. Stay tuned.

COMPARING SECONDARY CELLS   1.2.5

Table 1b compares features of commonly available rechargeable battery types.

Table 1b: Features for commonly available rechargeable cell types

Cell Type	Nominal Voltage	Electrolyte	Discharge to Zero Volts?	Cycle Life	Memory	Self-Discharge	Low Temp Performance	Peak Current
NiCd	1.2	KOH	Yes	High	Yes	Moderate	Good	High
NiMH	1.2	KOH	Not recommended	High	No	Fairly High	Good	Moderate to High
Lead-acid	2.0	Sulfuric acid	No!	Low-High	No	Moderate	Poor	High
Nickel Iron	1.4	KOH	Yes	Very High	No	High	Fair	Low
Lithium Ion	3+	Organic	No!	High	No	Low	Very Good	Moderate

SECONDARY BATTERY APPLICATIONS

ENGINE STARTING & SLI   1.3.1

Service: SLI stands for Starting/Lighting/Ignition – which is the ubiquitous automotive battery. Since these batteries are optimized for long float life, high peak currents, and infrequent shallow discharges, they would be a poor choice for powering an electric vehicle. They're designed with thin plates (for the high peak current) and either free or AGM (absorbed) electrolyte. They're not intended for cycling service.

As with all lead-acid batteries, high temperatures shorten their life.

EMERGENCY LIGHTING & ALARMS   1.3.2

Service: Long float; infrequent, moderate rate discharges (which may be shallow or full, but must be capable of 1.5 hours "run time" for emergency lighting as mandated by NFPA). They are not intended for cycling. Except for those infrequent discharges, they spend their entire lives on float charge at room temperature or above.

MOTIVE POWER   1.3.3

Service: Frequent full discharge and recharge cycling. In large enterprises, batteries may have two full cycles per day to satisfy the demands of three-shift operation. Their discharge profile ranges from low to high discharge rates. They *must* have high cycle life. They're usually flooded cells, except for batteries used in small trucks such as golf carts, which are usually gelled or absorbed electrolyte. Motive power batteries are usually used at room or slightly elevated temperatures. Subject to mechanical shocks during service and maintenance.

STATIONARY 1.3.4

Service: Long float life, with infrequent moderate-to-deep discharge at variable rates. This design needs a compromise between long float life and high cycle life. Designed for use in a permanent location.

Stationary batteries come in many flavors, from nickel-iron and NiCd to lead-acid of various constructions (flooded and sealed). There are also several standard bus voltages in use. Table 1c provides a quick rundown of the nominal number of cells needed for each bus voltage. They're based on 1.2 VPC for NiCd and 2.0 VPC for lead-acid. You may see differences, though, especially for NiCd, because of their high equalize voltage requirements.

Table 1c: Number of cells needed for each bus voltage by cell type

Battery Type	Number of Cells for Bus Voltage				
	12 v	24 v	48 v	130 v	260 v
Nickel-cadmium	10	20	40	100	200
Lead-acid	6	12	24	60	120

Remember that 1.2 VPC and 2.0 VPC are strictly nominal voltages, and that the open-circuit voltage of a lead-acid battery depends on its specific gravity. For more on this, see *Float & Equalize Voltages: How Do They Differ?* in Section 1.5.3.

DISCHARGING SECONDARY BATTERIES

THE CAPACITY OF a storage battery – that is, the useful amount of electricity that the battery can provide – is measured and expressed in ampere hours (Ah). A battery that can provide 10 amperes for eight hours is said to have a capacity of 80 Ah.

Of course, this is a fuzzy measurement. First, the rating assumes that you discharge the battery at a constant current of 10 A. That says nothing about the terminal voltage at the end of discharge. The following explanation uses a lead-acid battery as an example.

Stationary industrial batteries are usually rated at the eight-hour discharge rate, as in the example above. Capacities for automotive and other small batteries are frequently given at the 20-hour rate. To provide meaningful data for comparing batteries, manufacturers standardize the capacity tests by setting the end-of-discharge (EOD) voltage to a fixed value, usually 1.75 VPC (volts per cell). They may give supplemental data for discharges to other end voltages; 1.5 VPC and 1.8 VPC are common values.

Why different discharge voltages? It turns out that the capacity you can get from a battery, in Ah, depends on the discharge rate. The faster you discharge the battery (that is, the higher the current), the less capacity you can realize.

THE PEUKERT CONSTANT

If you're really interested in the nitty-gritty, here's a common standard form of Peukert's formula (from the Wikipedia site Peukert's Law):

$$t = H \left(\frac{C}{IH} \right)^k$$

If you know the Peukert constant for a cell, you can calculate the new discharge time, t, for any discharge current, I. C is the manufacturer's stated capacity in Ah, and H is the discharge time for that capacity (normally eight hours for stationary batteries). The exponent k is the Peukert constant.

If you don't know the Peukert constant, you can calculate it from any manufacturer's data sheet that includes figures for multiple discharge rates. There are many Peukert constant calculators on the internet.

As you can see from the text, the Peukert constant is of limited value for large stationary batteries and may actually be misleading.

This phenomenon was recognized early in battery history. The German scientist Wilhelm Peukert observed this behavior, and in 1897 developed an empirical formula to describe battery performance. [Peukert, pronounced *Poikert*, is relatively unknown, but his equation is very popular.] See the boxout *The Peukert Constant* for details.

Industrial battery manufacturers, however, haven't latched onto Peukert's potentially helpful math. There a few probable reasons for this:

- Batteries that are discharged at a high rate may be discharged to an EOD voltage lower than 1.75 VPC. Uninterruptible Power System (UPS) batteries, for example, are usually discharged to 1.5 VPC or lower, at rates from 1 minute to about 15 minutes. Conversely, when you discharge at a low rate, you should terminate the discharge at a terminal voltage higher than 1.75 VPC. The Peukert constant is based on discharging to the same end voltage regardless of the discharge rate. This makes Peukert's law difficult to apply to applications like UPS.

- The so-called *Peukert Constant* (for lead-acid, a number between 1.1 and about 1.4) isn't constant. A quick check of one battery spec sheet shows a range of 1.30 to 1.45 for discharge rates of 3 hours to ½ hour.
- The constant increases as the battery ages. A constant supplied by a manufacturer for a new battery will give unreliable results for a well-seasoned battery.
- The available discharge capacity is also a function of temperature, so a discharge test needs to adjust the resulting data for temperature. Peukert didn't explain how to do this.
- Peukert doesn't account for self-discharge at very low discharge rates.

So, industrial battery manufacturers' data include tables or graphs to help you determine usable capacity, without mention of Peukert. If you're building electric vehicles, though, people will Peukert you to death.

Three discharge methods are in common use to measure the Ah capacity of a battery: Constant Resistance, Constant Current, and Constant Power. In the real world, of course, batteries are rarely discharged at constant anything. We'll cover variable-rate discharges a little later, using an analysis procedure called Hoxie's method.

CONSTANT RESISTANCE DISCHARGE   1.4.1

A constant resistance discharge is the easiest to perform: just connect a fixed resistor to the battery and record the time it takes to reach the EOD voltage. It's normally used for small cells to provide a sense of their useful capacity for such loads as flashlights, music players, radios, and the like. As with all discharge tests, the time to EOD voltage is a function of temperature: discharge time is a little longer at elevated temperatures, and shorter – possibly much shorter – at low temperatures.

When a constant resistance discharge is performed, cell capacity is expressed in minutes (or hours) of operation, rather than Ah. This is because load current decreases as the cell discharges, making calculation of Ah cumbersome at best (at least without a computer). We compare constant resistance and constant current discharges in the next paragraph.

CONSTANT CURRENT DISCHARGE   1.4.2

There is no such thing as a constant current load in nature. Battery manufacturers must use special equipment to perform a constant current discharge. A constant current discharge is useful to rate battery capacity because it provides a simple and direct way to compare batteries and their performance at different discharge currents. A constant current discharge always results in a battery rating in ampere hours.

You know now that the Ah you can get from a battery decreases as you increase the discharge current. Manufacturers perform the tests by connecting a constant current load (electronically controlled) and monitoring the terminal voltage until the EOD is reached. If you use the same EOD voltage for all discharge tests, you can then compare McIntoshes with Winesaps.

But how do constant resistance and constant current discharges really compare? We can gain insight by examining what happens in small primary batteries. Consider the data for a high-quality, AA alkaline-manganese primary cell. For a constant current discharge of 22 mA, the service life, to 0.8 V EOD, is 100 hours, for a total of 2.2 Ah.

We'd like to determine the equivalent constant resistance discharge. We can try to find it for either of two cases: choose the starting current to be the same as the constant current discharge, or choose the final current, at the end of discharge, to be the same. If we choose R (the load resistor) to be 64 ohms, the initial current at the start of discharge (about 1.4 V) will be about 22 mA, the same as the constant current. This yields a service life of about 160 hours. But the slope of the voltage-time discharge curve isn't linear, so we don't know the actual average discharge current, or the Ah extracted from the cell. The cell takes longer to get to 0.8 volts, but we don't know if we actually get more out of it.

Another equivalent discharge is to use the manufacturer's data to determine the R value for a service life of 100 hours, which is 43 ohms. Now the initial current (at 1.4 V) is 32 mA, so it seems like we're getting more out of the cell, but the current drops as the voltage decreases, so we really don't know. Again.

If we assume that the terminal voltage decreases linearly with time (it doesn't, but this is only a test), we can estimate the Ah for the resistive discharge. The average voltage is 1.1 V, so the average current for a 64-ohm discharge is 17 mA, for a yield of 2.75 Ah. For the 43-ohm case, the average current is 25.6 mA, yielding 2.56 Ah. So we may, in fact, be getting more capacity for constant resistance than for constant current, which yields 2.2 Ah.

Note, however, that the manufacturer bases its tests on each discharge method supplying the same power at the end of discharge. This is the case where the final current would be the same for constant resistance and constant current, wherein the constant resistance discharge fares much worse. We get more, in the analysis above, only by tolerating lower power delivered near the end of discharge. Because of these differing test protocols, comparing the two discharge methods really isn't meaningful.

Let's look now at the constant-current discharge data for a large industrial battery; this example is a lead-acid battery rated 1,200 Ah to 1.75 VPC, with full-charge specific gravity of 1.240. Although the manufacturer doesn't provide a Peukert constant, we can calculate it from the discharge data.

Table 1d: Calculation of Peukert Constant from the variables in the other 3 columns

Discharge Rate	Current, A	Actual Ah	Peukert Constant
8 Hour	150	1,200	Baseline
3 Hour	330	990	1.24
2 Hour	435	870	1.30
1 Hour	668	668	1.39
0.5 Hour	930	465	1.52

We see from Table 1d that the calculated Peukert constant actually increases with the battery discharge rate. Using a Peukert constant based on a low discharge rate will give misleadingly high capacity ratings for higher discharge rates. For example, using 1.30 (the constant at the 2 hour rate) would give a discharge time of 1.4 hours instead of the actual 1 hour for a 668A discharge.

In summary, using the tabular data provided by the manufacturer is the only reliable way to predict capacity at any discharge rate, even though all the discharges are to the same end voltage.

TECH TIP

CONSTANT POWER DISCHARGE   1.4.3

Things get interesting now. While there aren't any naturally occurring constant power loads, there are a lot of applications where electronic equipment overcomes nature. Uninterruptible Power Supplies (UPS) are ubiquitous in modern industry, frequently used to maintain a computer's power during an emergency. If a device's power demand is constant, the UPS presents a constant power load to the battery during discharge. Data sheets for batteries designed for UPS frequently provide constant power (also called constant wattage) discharge data, often down to the one-minute rate.

In this case, it's possible to compare constant power and constant current discharges, as shown in Table 1e for a nominal 63 Ah battery (10-hour rate) discharged to 1.75 VPC.

Table 1e: Capacity at constant current, measured for different discharge rates. A six-cell, 63 Ah battery (10-hour rate), with 10.5 V EOD was discharged to 1.75 VPC. The table also compares constant current discharge to constant power discharge

Discharge Rate	Current, A	Power, W	Ah@ Constant Current
10 Hour	6.3	76	63.0
4 Hour	14.9	179	59.6
1 Hour	52.8	619	52.8
30 Minutes	93.5	1082	46.8
2 Minutes	404.1	4328	13.5

The battery is a six-cell unit, so the EOD voltage is 10.5 V.

Note that the available Ah is steeply related to the discharge rate, although not as extreme as in the battery used for the constant current example. This example is for a battery design optimized for high rate discharges. The significant point for these discharge data is that the power level in the constant power discharge is chosen to be equal to the *starting* power for the constant current discharge, which means that the currents are the same at the start of each discharge. The two discharge methods start at the same current, but the constant power discharge requires the current to increase as the discharge voltage drops.

As an example of these concepts, consider the discharge parameters for the 10-hour rate. The 76-watt discharge for the 10-hour rate has a starting current of 6.3 A, if the cell terminal voltage is 2.01 V, which is close to the open-circuit voltage. If the average discharge voltage is 11.25 V, then the average power delivered in the constant current discharge is 70.8 W, vs. 76 W in the constant power discharge. Using the same average voltage, the average current delivered during the constant power discharge is 6.75 Adc.

If you check the numbers on the higher discharge rates, you find that the starting power is reduced because the discharge voltage is less than 2.0 VPC at high discharge currents. This is due to the internal resistance of the battery; although low (only a few milliohms), at high currents it can reduce the discharge voltage of this battery by a volt or more.

KEY CONCEPT

EXAMPLE

WHERE DOES "UNUSED" CAPACITY GO?   1.4.4

By now, you are probably asking yourself, "If I can only get one-fourth of the capacity out of the battery at a high rate, where does the remaining capacity go?" The short answer is that it doesn't go anywhere; it's still in the battery. You just can't get to it.

In a lead-acid battery, one of the causes of reduced capacity is electrolyte mobility, or rather immobility, and the permeability of the active material in the positive and negative plates. By permeability we mean the ability of the electrolyte to diffuse through the active material and convert it to discharge current. During discharge, the sulfuric acid in the electrolyte is part of the chemical reaction. The acid reacts with lead or lead peroxide in the plates, reducing its concentration in the electrolyte, especially at the interfaces with the active material of the plates. The active materials, lead and lead peroxide, in turn, are transformed into lead sulfate. It's magic.

In other words, the sulfuric acid is turning into water. This reveals another cause of reduced capacity: the impedance (resistance) of the electrolyte is increasing, since water is a much poorer conductor of electricity than the sulfuric acid electrolyte. The increased resistance decreases the discharge terminal voltage even further. As the discharge continues, new sulfuric acid must migrate to the interface with the active material. This is the major electrochemical shortcoming of the lead-acid battery: the electrolyte must be involved in the chemical reaction.

KEY CONCEPT

At low discharge rates, the electrolyte concentration equalizes readily, and diffuses throughout the plates to contact most of the active material. At high discharge rates, however, the acid concentration cannot diffuse fast enough to make use of all the active material. The result is that the battery seems to be discharged (even though there is a significant amount of potentially useful active material left in the plates), and the terminal voltage starts to collapse. At some point, it is necessary to terminate the discharge to avoid permanent damage to the battery.

At very low discharge rates (the 20-hour rate or lower), you have the opposite effect. Most of the active material is available to the electrolyte, and the battery approaches full discharge at a fairly high discharge voltage. Therefore, you must terminate the discharge at a higher voltage than the "nominal" EOD voltage of 1.75 VPC. For low-rate applications, manufacturers generally provide data for discharges to 1.81 VPC.

At very high discharge rates, you can discharge a battery as low as 1.5 VPC. But always check with the battery manufacturer concerning your application, to be sure that your discharge rates and EOD voltages are optimized.

In the preceding sections, we've discussed idealized discharges: constant resistance, constant current, and constant power. In the real world, of course, battery loading is often variable, especially in utility applications. Complex discharge profiles have variations in current and depth of discharge. Later, we'll look at sizing a battery for these inconsistent loads.

HOW DO I SIZE A STATIONARY BATTERY?   1.4.5

Sizing for a Constant Load   1.4.5.1

Before you read about sizing batteries, be sure you've read *Discharging Secondary Batteries* (SECTION 1.4).

Although there are several steps, sizing a stationary battery for a constant load is fairly straightforward. Your first step is to determine the actual ampere hours that the battery must deliver to satisfy site or load requirements. If you have a constant current load, you're there: simply multiply the current by the operating time you need. Remember that at high rates, capacity is reduced, so check the discharge curves or tables for the battery type you want to use. But also, at high rates, you may be able to discharge to a lower EOD voltage.

This means that you already have a good idea of what battery you're going to use. Battery size is driven by the total Ah needed, so you can't necessarily trade off battery physical size (or cost) for discharge time. There may be surprises: after you've finished the calculations, you might find that the battery you selected won't do the job, requiring you to shift to a larger package.

For constant resistance or constant power, you need to convert the requirement into constant current, using any of the methods discussed above. Some guesswork is involved, so a good rule of thumb is to add another 5% or 10% as a design safety margin. For a constant power discharge, you could use the current at the end of discharge; this is the most conservative approach. Or you could use the constant power discharge curves, if the manufacturer provides them.

Once you have the Ah requirement, proceed as follows:

1. Determine the number of cells required. This is dictated by the upper and lower voltage limits for the site equipment, the charging voltage required for the chosen cell type, and the EOD voltage for the discharge rate.

TECH TIP

EXAMPLE

Example: A substation dc bus has an upper operating limit of 145 Vdc, and a lower limit of 105 Vdc. You've chosen a VRLA cell that requires a float voltage of 2.250 VPC at 25 °C, and the manufacturer recommends that the battery never be equalized. However, you know that the ambient temperature will go as low as 50 °F (10 °C), and you're using a temperature-compensated charger. The charger output voltage will be 2.325 VPC at 50 °F, temperature-compensated from the recommended 2.250 VPC at 77 °F. Therefore, the maximum number of cells allowed is 145 Vdc ÷ 2.325 VPC = 62.

For this application, EOD voltage should be no lower than 1.75 VPC. The minimum number of cells, if discharged to 1.75 VPC, is 105 ÷ 1.75 = 60. Sixty cells, of course, is the normal number for a (so-called) 130 V battery and is safe for the upper limit. We'll use 60 cells. In this case we're using the EOD voltage at 77 °F.

If you had a light load, with a long discharge time, say 24 hours or more, you would normally need a higher EOD voltage. Manufacturers publish discharge curves for an EOD of 1.81 volts. In this case, you could use 58 cells, and they would still meet your minimum system requirement of 105 V at the end of discharge.

For a battery that requires periodic equalization (a flooded cell, for example), you need to use the equalize voltage instead of the float voltage to calculate the maximum number of cells to meet the upper voltage limit. In the example above, a battery that requires an equalize voltage of 2.33 VPC at 77 °F would be equalized at 2.40 VPC at 50 °F. The maximum equalize voltage, 144.6 V, just squeaks under the maximum permitted bus voltage.

2. Adjust the Ah for the battery end of life capacity. Most manufacturers of lead-acid batteries rate the end of life as the point where the initial capacity is reduced to 80%. Therefore, multiply your Ah requirement by 1.25.

3. Adjust for temperatures below normal room temperature. If your battery is going to live forever in an environment of 25 °C (77 °F), lucky you. But a lot of applications are in unheated or poorly heated locations. The manufacturer's data sheet will give you a derating factor for temperature, or provide a graph or table of initial capacity vs. temperature.

Example: A lead-acid battery rated for 100 Ah at 25 °C might provide only 80 Ah at 0 °C. If your battery will experience significant time at this low temperature, multiply the Ah by 1.25. For the example above, with a low ambient of 10 °C, a correction of 1.1 would be adequate.

Don't adjust for temperatures over 25 °C, even though some spec sheets provide the data. The available capacity does increase at elevated temperatures, but that's gravy, added to your margin. But beware of sustained elevated temperatures. See *How Does Temperature Affect Battery Discharge?* (Section 1.4.6).

4. Now go back and check the curves for your selected cell type. It's possible that the corrections you've made will require you to select a larger cell. This isn't the time to cut corners. You don't want to get midnight telephone calls five years from now.

I assure you that I have no financial interest in industrial battery manufacturers. I'm just trying to help.

How do I size for a real-world load profile?   1.4.5.1.1

The information in the last section may be good for a UPS or similar application, which is close to constant power. But the chances are good that your world isn't like this. In utility and other stationary applications, the battery may be required to deliver widely varying load currents for an hour or several hours during a power emergency. The load profile for a dc system might look like Figure 1e during a five-hour backup period.

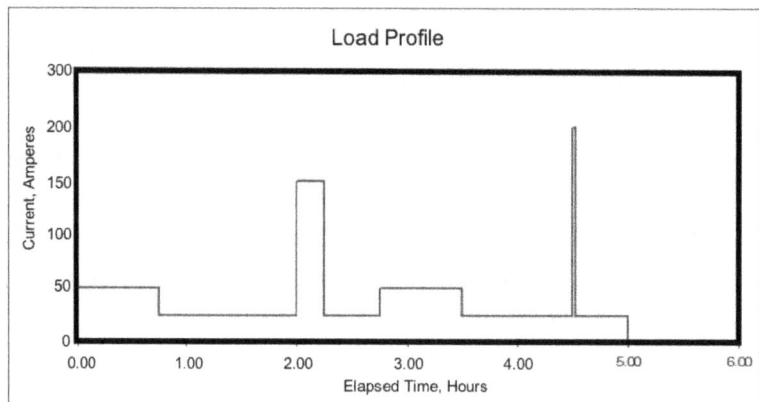

Figure 1e: Example of a real-world load profile that could be experienced by a battery during a power emergency

For this hypothetical discharge profile, the continuous standing load is 25 Adc, with intermittent loads of an additional 25 to 125 Adc. Near the end of the required backup time, there is a very short requirement for an additional 175 A. This transient load – known as a momentary load – may last only a few seconds, but we treat it as if it lasts for a full minute.

Why a full minute? Experience and testing show that even a very short but high-demand load reduces the battery voltage after just a few seconds; the reduced voltage means that the battery capacity is reduced even after the load is removed. While conservative, this is a good approach, especially if the demand is near the end of the discharge where the battery is nearly depleted.

If you add up all the load currents in the load profile shown above, you can calculate the total Ah removed from the battery. [In the example above, it's almost 200 Ah.] It seems that you should be able to size the battery based on those Ah. Remember, though, that higher current drains result in a lower capacity rating for the battery – this means that the battery needs to be sized larger than the calculated Ah would indicate, to be able to support the higher demands throughout the discharge profile.

TECH TIP

If it's not as simple as adding up the load currents, then how can you do it?   1.4.5.1.2

There's some history there, of course. In the 1950s, Earle A. Hoxie of the Electric Storage Battery Company of Philadelphia (Exide, among other brands), developed a method of calculating the battery size based on individual load currents, their discharge time increments, and the rated capacity of a single positive plate for a given battery type (over a range of discharge rates). This method was incorporated into IEEE Standard 485, "Sizing Lead-Acid Batteries for Stationary Applications." The method is complicated, cumbersome, iterative, and tedious. It's also very accurate.

Of course, you're going to keep me honest, so I used Hoxie's method to calculate the actual Ah requirements for the example above. The worksheet is shown in Appendix B. I used the example cell type given in IEEE Standard 485, knowing that you will check my work. I calculated for 1.75 VPC end voltage, using only room temperature. At the end, we find out that we need a 400 Ah battery to do the job in this example.

Naturally, to do a complete job, I would repeat the calculation for other plate sizes and battery types, until I found the most economical solution, or optimize for other requirements, such as floor space. I would also have to factor in the expected lowest operating temperature at the facility. Doing all this would be like mowing your lawn with a pair of scissors.

Is there a way to size the battery
without all that hassle?   1.4.5.1.3

Affirmative! Battery manufacturers provide computer software that does the whole job for you. Just specify the cell type you think will work and put in the currents and time periods for the discharge profile. The programs provide instant gratification, and allow you to run through many options quickly. Since most programs include a database of the manufacturer's cell types and their characteristics, you might not even have to preselect a cell type. Of course, the software won't give you the price; you'll still have to get that from the manufacturer.

Corrections   1.4.5.1.4

The sample calculation in Appendix B is uncorrected for the many contingencies surrounding battery applications. You will need to add a design margin and correct for temperature if your battery spends much time at cooler temperatures. You also need to add some margin for battery aging. These corrections are discussed in *Sizing for a Constant Load* in Section 1.4.5.1. The sizing programs will correct for temperature.

There is one more cause of reduced capacity. Brand-new batteries may not be fully formed; that is, they don't yet deliver their full rated capacity when they're shipped from the factory. It may take several months on float charge to attain that capacity. Normally, your corrections for design margin will compensate for that shortfall.

HOW DOES TEMPERATURE AFFECT BATTERY DISCHARGE?   1.4.6

In the earlier sections on primary batteries, and the general descriptions of secondary batteries, we noted that battery performance decreases at low temperatures, and is a little better at elevated temperatures. For a battery whose capacity is rated at 25 °C, capacity might be reduced to 70% at 0 °C. At 37 °C, you might gain 7% or 8% capacity.

There's more to the story, though. A lead-acid battery, operating at very low temperatures, say -20 °C, has significantly decreased capacity. The electrolyte at that temperature is less mobile and may freeze at the interface with the active material of the plates as the battery discharges and the specific gravity decreases. If ice crystals form inside the active material, the plates may be damaged. If you can avoid freezing, however, the capacity will recover when the battery returns to room temperature.

At the upper end, there is more risk of damage. At, say, 37 °C (body temperature), you gain a little discharge capacity, but remember that you're taking this out of the capacity rating at room temperature. In other words, you may be causing a deeper discharge than the battery is rated for at room temperature. If you discharge to 1.75 VPC at 37 °C, it's like discharging an extra 5% at room temperature. In addition, battery life decreases at elevated temperatures: at 37 °C, the battery life will be less than half of the rated life at room temperature.

For more in-depth information on temperature effects, see *Temperature Effects* in CHAPTER 5.

CELL REVERSAL: HOW DOES IT HAPPEN? WHY AVOID IT?   1.4.7

We mentioned cell reversal in discussing secondary batteries in general. Cell reversal can occur in any series string during discharge. Cell reversal is bad. We picture the process in Figure 1f.

Figure 1f: Cell reversal process illustrated progressively with a three-cell string (for simplicity): normal state where all cells are fully charged is on top; in the middle diagram, the middle cell has fully discharged; in the bottom diagram, this cell has reversed polarity, further lowering the voltage across the load, & causing the other cells in the string to try to charge it. Courtesy Tony Ortiz, HindlePower, Inc.

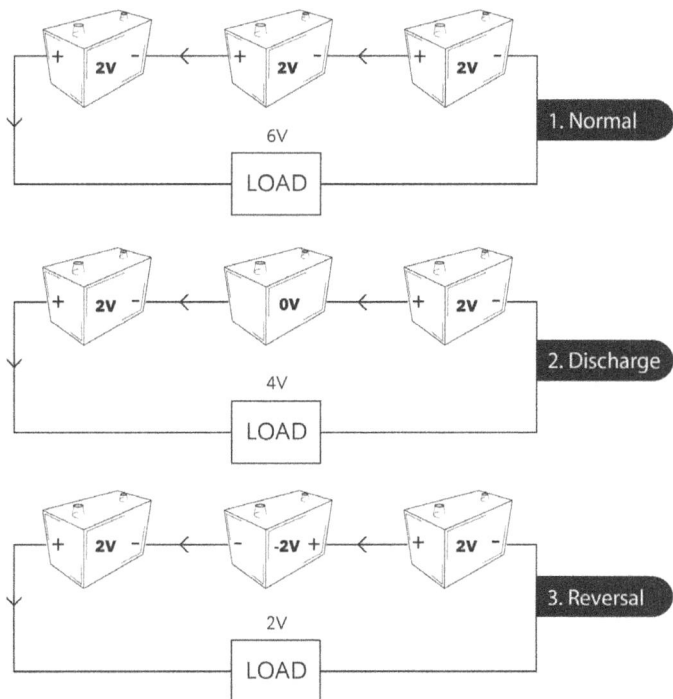

Due to manufacturing tolerances, differences in aging, and cell imbalances resulting from frequent cycling, some cells in a series string will have less discharge capacity than others. If you're discharging a battery to a very low voltage, it's possible that a cell with significantly lower capacity will become fully discharged while the remaining cells still have a way to go. In a string of, say, 60 cells, the loss of one 2 V cell might be tolerated by the load, and the system wouldn't send an alarm.

Now what happens to that discharged cell? It's no longer contributing to the load current, but load current flows through it – only now, it's in reverse. That is, the rest of the battery is trying to charge the weak cell in reverse, turning the anode into a cathode and vice versa.

Not good! As you have read, the positive and negative plates in a lead-acid battery are designed differently, each optimized for its intended use. During discharge, the active material in each plate converts into lead sulfate. Trying to convert the active material into its opposite polarity is not only futile but causes permanent damage. It's possible that normal charging cannot later recover the cell.

The capacity of a cell whose polarity is reversed is small, and the cell will gas after only a few minutes, producing an explosive mixture of hydrogen and oxygen, and consuming electrolyte. This causes permanent capacity loss in a VRLA cell.

You can avoid cell reversal by paying strict attention to EOD voltages. A lower limit of 1.5 VPC should be a safe cutoff voltage for large battery strings at high discharge rates; use 1.75 VPC or higher for moderate to low discharge rates.

SAFE PRACTICE

»   *Q: We recently had an extended power failure, and weren't able to shut down the dc loads. Our battery went flat. I mean zilch. Can we recover anything?*

A: *You're in a difficult situation. When you finally were able to disconnect the load, was there any voltage at all on the battery?*

» Q: **It went up to about 80 volts (for a string of 60 cells).**

A: *OK. That's an average of about 1.3 VPC. When your open-circuit voltage is less than about 1.8 VPC, the battery is in real trouble.*

» Q: **Why is that?**

A: *During discharge, the electrolyte is gradually changing to water. Remember that the open circuit of a lead-acid cell is, roughly, 0.85 plus the specific gravity of the electrolyte. When you have a fully-charged cell with a specific gravity of, say, 1.215 (pronounced twelve fifteen), the open-circuit voltage is 2.06 V. If the battery were filled with pure water, you'd have no voltage at all: no electrolyte, no voltage.*

But an open-circuit voltage of 1.3 VPC means that there's some electrolyte left, and some non-discharged active material. You can't really calculate the specific gravity, since the average would be less than 1.0, an impossibility. Since some of the cells are probably reversed, that means that some have a higher open-circuit voltage, and therefore some measurable specific gravity.

» Q: **You're getting very technical. What's the bottom line?**

A: *The battery is probably toast. But you may be able to recover a little capacity, which could tide you through small emergencies until your new battery arrives. Call your salesperson.*

» Q: **So what's the lowest open-circuit voltage that I should ever see?**

A: *No lower than 2.0 VPC. That means a specific gravity in the discharged state of 1.150 (eleven fifty), which would keep the battery from freezing down below 0 °C. Before you try to measure the open-circuit voltage, remember to let the battery rest for a while,*

TECH TIP

with no load or charger connected. It will appear to be lower at the end of a discharge, and it will start to recover within a few minutes after the load is removed.

HOW DO I PROTECT A DC SYSTEM FROM SHORT CIRCUITS?   1.4.8

Despite all our best efforts to avoid catastrophes, it's possible that equipment failure, human error, or a misalignment of the stars might cause a battery to be short-circuited. This is not a good thing.

TECH TIP

It's a good idea to plan for this eventuality. A dc system should be designed so that a fault in any connected equipment will activate circuit protection for the branch that is faulted. This will protect the battery and maintain the operation of the rest of the dc system. Usually, some last-ditch circuit protection is provided for the battery also. If all the protective devices are coordinated properly, branch circuit protection will always activate before the main battery protection activates, which would leave the entire system in the dark.

TECH TIP

So, how much short-circuit current should we protect against? A common rule of thumb is that a flooded lead-acid battery can deliver about ten times its one-minute discharge rating for anywhere from a few milliseconds to several seconds. As the fault continues, the battery impedance rises quickly, so the short-circuit current will decrease, although not to a safe level. Some cell designs, such as VRLAs designed for high-rate discharges, can produce higher instantaneous fault currents, but the current may fall off quickly because of the electrolyte immobility. If the manufacturer's data sheet doesn't give the one-minute discharge rate, you can use the nominal Ah rating, which would provide a good approximation (that is, the short-circuit current would be ten times the stated Ah rating). Some data sheets provide the short-circuit current rating.

So what happens if branch circuit protection fails? Of course, a fuse somewhere will clear the overcurrent, protecting the battery from permanent damage. And of course, New York City will go dark. Again. But what if a fuse doesn't clear? Or what if the system designer simply omitted protection?

System faults, of course, have some impedance. There is the internal resistance of the battery itself, and the impedances of the wiring to the dc bus, the branch circuits, the fuses that aren't blowing, and so forth. Plus, the faulted equipment might still support a little voltage. The higher the system impedances, the lower the battery fault current will be. But we should assume that the current, if sustained, will be high enough to melt internal battery components, causing permanent damage, and possibly fire, or reverse several cells in the battery.

TECH TIP

Battery manufacturers' spec sheets may give you values of battery short-circuit current, or internal resistance, or both. You can use these values to design the last-ditch protection. Also, you can consult IEEE Std. 1375, "Guide for the Protection of Stationary Battery Systems," which will tell you everything you need to know and a lot more. The standard includes a procedure for calculating cell internal resistance. Appendix B has a sample calculation of cell internal resistance.

One more thing to remember: the values for short-circuit current are for a fully charged lead-acid battery. As a battery discharges, its internal impedance rises. If a fault occurs toward the end of a battery discharge, the available short-circuit current may be reduced by up to 50%. This isn't true for NiCd or NiMH batteries: since the specific gravity doesn't change for these systems during discharge, the internal resistance is more constant, and the available short-circuit current at the end of discharge is closer to the value for a fully charged battery.

CHARGING SECONDARY BATTERIES

WE CHARGE (OR recharge) secondary batteries by applying direct current to the terminals, in the opposite direction from discharge. This is probably not news. The charging process needs to be carefully controlled to avoid undercharge or overcharge, either of which could damage a battery in the long term. We also need to limit the starting current for a recharge to a level that is safe for the battery and the charging equipment, but that minimizes the charging time. Sounds like we're looking for a compromise.

When we discharge a battery, we're taking out charge – expressed in ampere hours. Fully charging the battery requires that we replace the ampere hours, but because the charging process isn't 100% efficient, we need to put in a little more. Generally, lead-acid batteries require about 110% of the ampere hours removed, and NiCd or NiMH batteries require 120% to 130%.

Charging methods can be constant current or constant voltage. Constant current charging is the fastest way to return ampere hours to the battery and is usually used for motive power batteries. Once the battery is charged to a specified voltage (about 2.3 VPC for lead-acid), the charging current is reduced to a finishing rate for a specified time, or the charger switches to constant voltage for a maintenance charge. Or both.

Stationary batteries cannot use constant current because there are upper voltage limits to connected equipment. Constant voltage charging is used, with the charger sized to deliver charging current of about 10% to 20% of the battery Ah rating, while still providing power for the load on the dc bus. Two voltage levels are generally used: float, for long-term operation, and a higher voltage for equalizing cells when necessary.

In either case, charging must be carefully controlled near the finish to minimize potential damage from excessive gassing, while ensuring that the battery reaches full charge. Voltage settings must also account for operating temperatures above or below room temperature.

Read on for more details. Except where noted, the material below pertains to lead-acid batteries.

THE AMPERE-HOUR LAW   1.5.1

In 1925, George Wood Vinal described the ampere-hour law for lead-acid batteries (actually articulated by J. L. Woodbridge in earlier research). It states, in essence, that a lead-acid cell will safely accept a charge current (at any arbitrary voltage) that is equal to the ampere hours of capacity remaining to be charged. By "safely," we mean that there will be no excessive gassing or temperature rise in the cell. For example, a battery that has been discharged by 200 Ah can accept an initial charge current of 200 A, without danger of excessive gassing or temperature rise.

The implication of this law is that a completely discharged cell will accept a large charge current, but that the charge current must be progressively reduced to meet the "Ah remaining" conditions to avoid excessive gassing. This is an empirical law, based on experiments performed as early as the 1920s, when all lead-acid batteries were of flooded construction and some gassing was an expected part of the charging process. But the requirement for continuously decreasing current makes constant current charging for modern (especially VRLA) lead-acid batteries impractical, and possibly damaging. Fortunately, constant-voltage charging (properly done) can eliminate the risk of excessive gassing, and temperature-compensated charging reduces the risk of excessive temperature rise.

There are, of course, practical limits on charge current, due to the finite capabilities of internal battery connections, charger components, and the ac power grid. Also, it's important to understand what we mean by "fully discharged" and "fully charged." Check the boxout *What "Fully" Means* for those definitions.

WHAT "FULLY" MEANS

When a discharge cycle is finished, for any reason, you can (almost) always take more out of the battery. But should you?

For our analysis here, we consider a lead-acid battery to be fully discharged when any further discharge would violate any of the limiting conditions that we've defined, such as the EOD voltage or the discharge specific gravity.

Defining fully charged is trickier. Strictly speaking, full charge is achieved when the specific gravity of the electrolyte is restored to the manufacturer's initial specification. But the full-charge specific gravity decreases as the battery ages, and active material is lost to sulfation and corrosion.

From the user's point of view, full charge could be defined as when the battery can repeat the previous discharge. This requires replacing the Ah discharged with about 110% of that value for a lead-acid battery.

CHARGE ACCEPTANCE   1.5.2

What is charge?   1.5.2.1

We know from the ampere-hour law that a discharged battery will accept charge very rapidly at the start of the charging cycle. But what do we mean by charge?

Unless you're Reddy Kilowatt®, you can't hold electricity in your hand. But imagine that you could: cup your hand and imagine a puddle of electricity in it. Now tilt your hand, and the electricity will start to pour out in a thin stream. That stream is discharge current. As it streams out, it reduces the amount of electricity remaining in that puddle in your hand. That puddle is charge.

If you could reverse gravity, you could also get that stream to reverse, and trickle back up into your hand, increasing the amount of electricity (charge) in the puddle. Now the upward stream is charging current, instead of discharging current.

You can see that the longer you let the stream run, the more charge will run out of (or into) your hand. Eventually, your hand will be empty, or full. Likewise, if you tilt your hand a little more, and the volume of the stream increases, your hand will empty sooner. The charge in your hand is limited by how much electricity you could hold at one time.

Now suppose you're using your anti-gravity tool and charging current is streaming up into your hand. Your hand starts to get full, and some of the current starts to leak out at the edges; you'd like to hold it all, but some escapes anyway. Once your hand has all the charge it can hold, the rest just leaks out.

Sounds silly, right? But it's essentially what happens in the battery as you reach full charge. The charging current won't actually "leak out," but it doesn't add to the total charge in the battery, which means it must have some other effect.

What happens to the charge that isn't accepted?   1.5.2.2

First, what happens to the charge that *is* accepted? At the start of charging, virtually all the current is used to convert the discharged active material back into its charged state. That is, lead sulfate in the negative plate is converted back into lead, and lead sulfate in the positive plate is converted back into lead peroxide. [If you like chemistry, you can check the equations in Section 1.2.2, *Lead-Acid Batteries*.]

You can see that the longer you pump charging current into the battery, the greater the charge will be. In geek talk, *charge* is the time integral of current. Just so you know, one ampere flowing for one second will give you one *coulomb* of charge, and an ampere flowing for one hour gives you, of course, one ampere hour. So you can see that one Ah is equal to 3,600 coulombs. See the boxout *Power & Energy in a Lead-Acid Cell* to learn how much power is expended in charging a 2V battery with one Ah.

POWER & ENERGY IN A LEAD-ACID CELL

In dc circuits, power = voltage x current. Power is expressed in watts, which equals volts x amps.

Therefore, if we charge a 2 volt battery with 1 Ah of charge, we have:

2 volts x 1Ah = (2 volts x 1 amp) x 1 hour = 2 watts x 1 hour = 2 watthours, or 7200 watt-seconds.

Since a watt-second is a joule, the unit of energy:

2 volts x 1Ah = 7200 joules.

From this example, we see that Energy = Power x Time; in geek speak, energy is the time integral of power.

But at about 80% state of charge (SOC), charge acceptance drops below 100%. So once we've put back about 80% of the ampere hours that we took out during the previous discharge, the charging action isn't 100% efficient, and we need to put back a little more than 100% to reach full charge. We normally use a factor of 1.10 for lead-acid batteries. This means that to recharge 100 Ah, we need to put in 110 Ah. The inefficiency of the charging process in this region means that the excess energy is converted to heat.

Electrolysis and Hydrogen Evolution   1.5.2.3

At 80% SOC, we're also bumping up against other constraints. Most of the active material has already been converted, and the remaining material isn't as accessible to the electrolyte and the charging current as the first 80%. It's more difficult for the charging current to reach the active material and electrolyte deep in the plate's pores. Some of the energy, then, starts to overcharge the already charged material, leading to electrolysis. This doesn't add to the charge in the battery; instead, it causes gassing, water loss, increased maintenance requirements, grid corrosion, and active material loss in the positive plates. Charging wears out a battery and shortens its life.

And yet, some gassing is an inevitable consequence of fully charging a battery, and in a flooded cell design, has the benefit of mixing the electrolyte, reducing stratification (Section 1.5.2.5), and homogenizing the specific gravity of the electrolyte.

Battery Room Ventilation   1.5.2.3.1

Gassing also releases hydrogen and oxygen, as we described in *Lead-Acid Batteries* in Section 1.2.2. If a flooded battery is installed inside a building, it's important to provide an adequate air exchange to control the concentration of hydrogen well below explosive levels. Manufacturers' application manuals provide formulas for calculating the amount of air required, based on battery type, size and age, float voltage or float current, and the number of cells. The manuals generally provide additional techniques for estimating the amount of water that will need to be replaced because of electrolysis in normal operation. More frequent watering may be required in applications with elevated temperatures.

One more thing: in a flooded cell with antimony alloy grids, gassing also releases stibine, a gaseous compound of antimony (antimony hydride). Stibine is poisonous, and

SAFE PRACTICE

in extreme cases can cause kidney failure. Ventilation adequate for hydrogen evolution also protects against stibine build-up.

Specific Gravity   1.5.2.4

Now consider specific gravity. What's happening with that electrolyte during charging? In *Discharging Secondary Batteries* (Section 1.4), we pointed out that the specific gravity decreases during discharge, basically turning to water. In the manufacturer's spec sheet, the specific gravity is usually given, but that's for a new, fully-charged battery at room temperature, with Venus visible at sunset.

At the end of the discharge, the electrolyte is depleted, and is a poor conductor. During charging, the specific gravity increases, and the battery impedance decreases. As the specific gravity approaches its nominal value, it becomes more difficult to get in those last few ampere hours. While we talk about an eight-hour recharge, it may take a week or more on float charging to bring a stationary battery to full charge. Motive power batteries, because they're usually charged by a two-rate constant current, may get closer to full charge in an eight-hour shift, but at the expense of more gassing.

Battery manufacturers recommend that the specific gravity of each cell be checked during periodic maintenance. These measurements, made with a *hydrometer*, should be performed by a trained technician. The readings must be corrected for temperature, since the nominal specific gravity is expressed at room temperature.

A note about NiCd and NiMH batteries: In these systems, the electrolyte (KOH) acts only as a medium for exchanging ions, and it's the ion exchange that drives the external current. The specific gravity doesn't change during discharge or subsequent recharge, and so the internal impedance is more constant than in a lead-acid battery. You can see this in the flatter discharge curve of voltage vs. time.

Electrolyte Stratification   1.5.2.5

Back to lead-acid batteries for a moment. In flooded cells, the electrolyte would ideally be evenly distributed throughout the cell. That is, the specific gravity would be the same, no matter where you measure it. But the electrolyte in a flooded cell has high mobility and, to boot, sulfuric acid is heavier than water. In a cell that has been on float charge for an extended period (or off charge completely), the higher density acid tends to sink to the bottom of the cell, leaving a weaker electrolyte at the top of the cell. Higher acid concentration at the bottoms of the plates can cause excessive sulfation of the grids.

During a discharge, the active material at the bottom of the cell provides more of the discharge current than the stuff at the top of the cell. The result is a plate that is unevenly discharged from top to bottom. On subsequent recharge, the tops of the plates will be overcharged before the bottoms of the plates are fully recharged, causing more grid corrosion and all the other nasty things that shorten battery life.

How to overcome this? Of course, you could pick up the cell and shake it, but I'm not applying for that job. Some large cells are equipped with stirring or aeration mechanisms to keep the electrolyte circulating. Equalize charging, and the gassing that results, is a good way to stir the electrolyte. Of course, this leads to increased maintenance.

By the way, the shaking method is more or less automatic in your car battery if you have a flooded design. If not, don't worry: VRLA batteries are far less prone to stratification, since the electrolyte is absorbed or gelled, resulting in lower mobility.

Sulfation   1.5.2.6

When a lead-acid battery is discharged, the active materials are converted to lead sulfate, of a form that can be re-converted to the original active material during recharging in normal use. There is always a speck, though, that doesn't get reconverted. This material eventually grows into a larger crystal structure, which resists charging. Over time, and many recharge cycles, the amount of crystal growth increases, resulting in grid growth; this robs the battery of useful active material, and therefore capacity. It also reduces the specific gravity over time.

Sulfation can be accelerated if a battery is left in a discharged state because the crystal growth is a slow but continuous process. Unlike NiCd cells, lead-acid cells cannot be stored in a discharged state, and in fact must be periodically refreshed when in storage.

Grid Corrosion   1.5.2.7

This occurs during the final phases of charging, when some of the charging current is going into electrolysis. The problem is chiefly at the positive plate, where oxygen is evolved; the oxygen reacts with the grid and active materials, causing the active material to lose contact with the grid, reducing capacity and increasing internal resistance.

In VRLA cells, the evolved oxygen and hydrogen are encouraged to recombine, replacing the water lost by electrolysis. However, there is always some corrosion, with attendant loss of electrolyte, and this is one of the limitations to the life of a VRLA cell.

Active Material Shedding   1.5.2.8

During discharging of a flat plate (flooded) cell, the plate enlarges because lead sulfate takes up a little more room than lead peroxide. Conversely, recharging causes

Figure 1g: "Positive active mass fractioning" in article from authors Petr Krivik & Petr Baca called "Electrochemical Energy Storage" found in Chapter 3 of "Energy Storage – Technologies and Applications" by Ahmed Faheem Zobaa, ISBN 978-953-51-0951-8, published: January 23, 2013 under CC BY 3.0

the plate to contract again. This mechanical action gradually causes active material to be loosened from the plate, as pictured in Figure 1g. After it loosens, the material settles to the bottom of the cell jar.

Flooded cells are constructed with a space under the plates for the shed material to collect. During the cell's lifetime, the material gradually fills this space, and eventually may contact the plates above the space, shorting the battery. This prevents the cell from taking a charge, so that it must be serviced or replaced. At this point, the battery might have reached its end of life anyway, since serious shedding is one sign of frequent deep cycling.

FLOAT & EQUALIZE VOLTAGES: HOW DO THEY DIFFER?   1.5.3

Remember specific gravity? In lead-acid cells, the specific gravity determines the open-circuit voltage of the cell: the voltage is 0.85[4] plus the specific gravity. For example, a cell with a fully-charged specific gravity of 1.215 has an open-circuit voltage of 2.065 volts.

---

4   Actually, 0.845. Most battery instruction manuals use either 0.84 or 0.85.

Why do batteries need to be recharged regularly, even if they aren't used?   1.5.3.1

Like any electrochemical system, though, a lead-acid cell can't just sit on a shelf and hold its charge forever. A lead-acid cell self-discharges through chemical action between the electrolyte and active materials, even though it's delivering no external current. A cell in storage needs to have a refreshing charge at intervals no longer than about six months; if a cell is allowed to self-discharge for a longer period, some active material, and therefore cell Ah capacity, will be irretrievably lost.

As the cell self-discharges, the specific gravity decreases, and the open-circuit voltage decreases along with it. So, we connect the cell to a charger and restore it to full charge. We know when we're done because the specific gravity will be back at the proper value. If we can't measure the specific gravity, we can go by the open-circuit voltage. This freshening charge is usually done at about 2.4 VPC.

TECH TIP

What about just keeping a battery connected to a charger permanently?   1.5.3.2

Stationary batteries are normally permanently connected to a charger, which maintains the charge on the battery while also providing power to the connected dc loads. The charger is adjusted to a float voltage that just compensates for the self-discharge of the battery. The float voltage must be within the acceptable voltage range of all the equipment on the dc bus. [See also SECTION 1.4.5, *How Do I Size A Stationary Battery?*]

How do we know the proper float voltage?   1.5.3.3

In general, the manufacturer's spec sheet provides this information. If this isn't available, but you know the nominal specific gravity, you can calculate the open-circuit voltage. An acceptable float voltage would be 1.05 times the open-circuit voltage for lead-antimony, or about 1.08 times for lead-calcium.

Don't know the specific gravity, either? You're on thin ice, but you can try the following technique: Once you are sure the battery is fully charged, measure the float current while you adjust the float voltage. Try to get the float current within these ranges:

Lead-antimony	40 ± 20 mA per 100 Ah
Lead-calcium	11 ± 5 mA per 100 Ah
Lead-selenium	20 ± 10 mA per 100 Ah

These numbers are for room temperature, with the battery capacity given for the eight-hour rate. And, of course, the battery must be healthy, and nearly new, since float current rises gradually as the battery ages.

How long can stationary batteries remain on float charging without problems?   1.5.3.4

Stationary batteries can remain on float charging for a very long time without significant degradation and perform admirably at the next power emergency. There is a fly in the ointment, though (isn't there always?). Maybe two flies. The first is that multiple cells connected in series, when on float for long periods, may become mismatched in capacity. This would be evidenced by differences in specific gravity (or open-circuit voltage) from cell to cell.

The second fly shows up when a battery is frequently cycled. Normally, you don't want to do this with stationary batteries, especially calcium alloy, but it happens. With repeated discharge-charge cycles, some cells won't be returned to full charge as soon as others; this may cause some cells to be overcharged more, causing more gassing.

What is equalization charge? How does it help prevent these problems?   1.5.3.5

The cure is the same in both cases: an equalization charge. This is a charge at voltage higher than float for a fixed period. The intent is to bring the "weaker" cells up to full charge so that they are matched with the stronger cells.

Note, though, that the terms "weaker" and "stronger" don't refer directly to cell capacity or quality, but to the on-charge voltage for the cells.

TECH TIP

You're saying, "But won't that overcharge the stronger cells even more?" Well, yes. And it's the reason that we normally equalize only flooded batteries, so that we can replenish the water that is inevitably consumed. VRLA batteries normally don't need equalization, and most manufacturers recommend against it. It's also important to terminate the equalize charge after a reasonable time; spec sheets usually recommend 12 to 24 hours. Also, equalize charging should be infrequent, preferably only when it's indicated by imbalanced specific gravity or voltage readings. In some installations, equalize charging is initiated automatically after a power failure, when the ac power is restored. This automatic operation is predicated on the theory that the cells become mismatched during the previous discharge. For more on *Equalization Charging*, see Section 1.5.5.

» *Myth: Keeping the charger set to equalize increases the capacity of the battery.*

  Whoa! Equalizing a battery maximizes the capacity on the next discharge because it brings "weak" cells up to full capacity. But it can't increase the capacity of any cell beyond its design value. The capacity of a battery, remember, is determined by its physical attributes, including such factors as grid alloy, plate design, specific gravity, discharge rate, and temperature. As soon as you start a discharge, the terminal voltage drops quickly to the loaded value, regardless of whether you start at the float or equalize voltage.

Keeping the charger set to equalize will accomplish one thing: it will wear out the battery sooner.

CHARGE PROFILES: WHAT ARE COMMON METHODS FOR RECHARGING?   1.5.4

Goals   1.5.4.1

The goals of any charge profile are obvious: recharge the battery to full capacity as quickly as possible and make the battery last forever. Oh, and make sure you don't overtax the charger components, or the dc bus wiring, or the ac power source, and keep the charging voltage (for stationary batteries) within spec for the connected loads. There are several approaches, depending on the battery application.

Motive Power: I-V-I Charge Profile   1.5.4.1.1

Motive power batteries are generally recharged with a constant current profile, at least for the start of charging. A popular profile is called the I-V-I (or I-U-I in Europe): a starting constant current that can be as high as C/2 (where C is the Ah rating of the battery), followed by a constant voltage period, and finishing with another constant current at a low rate.

The changeover from constant current to constant voltage is triggered by the battery reaching a critical voltage, usually 2.3 to 2.4 VPC, just below the onset of "hard" gassing. If the starting current is chosen correctly (about the C/2 to C/4 rate), this occurs at roughly

80% state of charge. During the constant voltage period, which may be set at about 2.4 VPC, the current gradually decreases until it reaches the finishing current, which can be set to C/10 to C/20. At that point, the charger reverts to constant current at the low rate and allows the voltage to rise without limit.

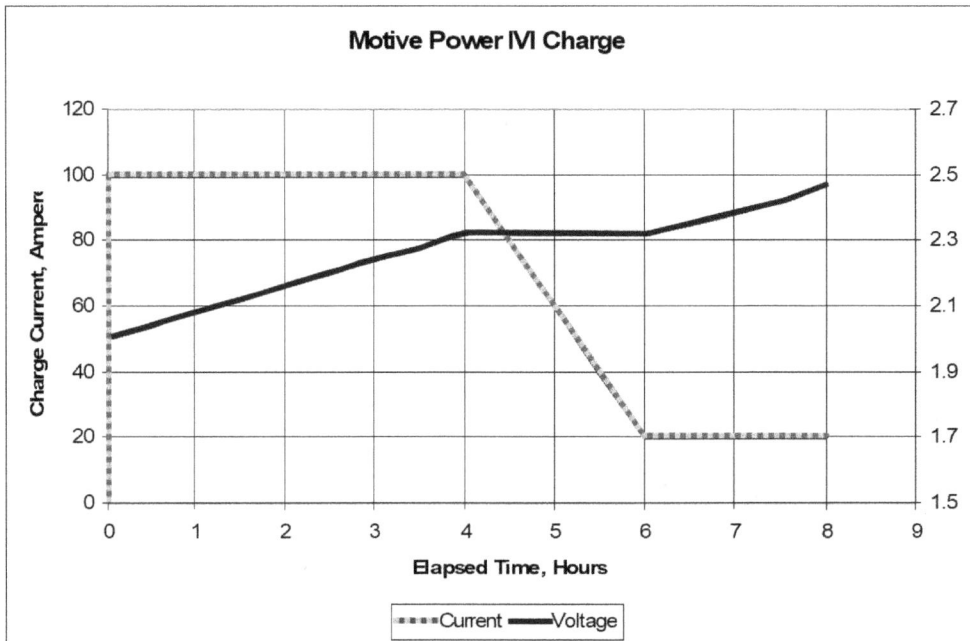

Figure 1h: Profile typically used to charge motive power cells called I-V-I because it begins with constant current (I) followed by constant voltage (V) and constant current

Normally, a timer terminates the charging. The timer may be started at the start of the charging cycle or be triggered by the switchover to the finishing rate. In the latter case, there should be an override timer, in case the current doesn't decrease to the finish rate (and start the timer), which could be caused, for example, by a shorted cell.

A graph of this profile is shown in Figure 1h.

Stationary: Constant Voltage   1.5.4.1.2

Stationary batteries are recharged at constant voltage. Both the float and equalize voltages are chosen to be within the acceptable range for the dc equipment connected to the battery. This means that specifying the float and equalize voltages is an integral part of sizing the battery, as is choosing the correct number of cells for the application (See *How Do I Size a Stationary Battery*, SECTION 1.4.5).

KEY CONCEPT

Normally, a charger is sized to recharge the battery in eight hours, while still powering the total connected load. This is fine in applications where the load is constant. In some applications, however, the load isn't connected until there is an ac power failure, or the load is intermittent during normal operation. In a substation, for example, the charger is usually sized to recharge the battery without a dc load connected, or with very small load current.

Float chargers incorporate output current limiting (what's that? See Chapter 4 *Output Current Limit*). When we first start to recharge a discharged battery, the charger is limited to a current that is usually 100% to 110% of the *charger* rating (not the battery rating). In a substation, for example, a charger might be rated to charge in eight hours, meaning that it would start at about the C/6 rate. Starting at the C/8 rate wouldn't restore enough charge in eight hours because the charge acceptance for the last 10% or 20% isn't as good as for the first 80%.

Example: A substation has a 75 Ah battery (eight-hour rate) that has been fully discharged, and a charger rated at 12 Adc output. Charger current limit is set to 13.2 A (110% of rating). In the first six hours of charging (if the charger operates in current limit for the entire time), the charger delivers 79 Ah to the battery, which is enough to bring the battery to 80% to 90% state of charge. At that point, the current starts to decrease because the output is now limited by the float-voltage setting.

EQUALIZATION CHARGING   1.5.5

In Sections 1.5.3.4 and 1.5.2.5, respectively, we described the need of a flooded lead-acid battery for occasional equalization – a recharge cycle at a voltage slightly higher than float voltage – to re-balance cell capacities and correct any electrolyte stratification.

What constraints must I consider when equalizing stationary batteries? 1.5.5.1

Voltage Limits on Connected Equipment 1.5.5.1.1

Electrical and electronic equipment powered by a dc bus has a power supply voltage "window" for safe and reliable operation. Equipment on a 24 Vdc bus may have an upper supply voltage limit of 30 Vdc; on a 125 Vdc bus, the upper limit is usually 150 Vdc.

A dc system design that includes equalization must keep these voltage limits in mind. The maximum equalize voltage for the battery must be less than the lowest upper-voltage limit for any connected equipment. This dictates the maximum number of cells allowed for the chosen cell type, based on the cell's required equalization voltage. For more on this, see *How Do I Size a Stationary Battery?* in Section 1.4.5. If you're using a temperature-compensated charger, you need to know the maximum voltage at the battery's lowest operating temperature at your facility.

If you have a VRLA that doesn't require equalization, it's a good idea to disable the equalization capability of the charger or ensure that the equalize voltage is set to the same value as the float voltage.

In the previous section on charge profiles, we described float charging for stationary batteries. When a charging profile (sometimes called a *charge regime*) includes equalization, it is usually a two-voltage profile, with the higher equalize voltage being held for some time. The equalize time can be fixed, or be proportional to a battery charging characteristic, such as the time required for the on-charge voltage to reach a critical point, as in the profile for motive power batteries.

End Cells & CEMF Devices 1.5.5.1.2

Even with careful planning and cell selection, there may be applications where equalization is required, and the resulting dc voltage would exceed the load's maximum specs. One solution is to disconnect the load while equalizing; this is rarely a viable option but may be possible in installations with redundant batteries.

A time-tested method to allow high voltage equalization is to use "Counter EMF," or CEMF cells. Historically, one or two extra electrochemical cells were inserted between the battery and the load; the cells were usually unformed alkaline cells, meaning that the active material had nearly zero capacity. When the charger was set to the equalize mode, these cells provided a voltage drop between the battery and the load to hold the voltage on the load below the maximum allowed value. During float charging, or during discharge, the cells were shorted out with a contactor.

This method is like the "End Cell" concept, originally used in telephony, which inserted low-capacity cells in series during battery discharge to maintain usable bus voltage.

The obvious disadvantage to CEMF cells is that they increase maintenance requirements, since they use a liquid electrolyte which must be replenished. Electrolyte is lost during equalization charging because CEMF cells gas copiously. Modern CEMF devices use silicon diodes in place of electrochemical cells, and automatically-controlled switching devices to insert or remove the devices from the circuit.

How do I equalize if I have chargers connected in parallel?   1.5.5.2

If you have two (or more) chargers connected in parallel, you have a choice of two modes of operation: random load sharing and forced load sharing. More information on parallel charger operation is found in the first section of CHAPTER 7.

If you are operating two chargers with forced load sharing, both chargers must be set to the same operating mode – float or equalize – for the output load current to be shared. With load sharing enabled on some chargers, such as HindlePower's AT series charger, when you set the primary charger to equalize; the secondary charger is controlled by the primary, so that both remain in equalize mode until the end of the equalize charge.

One caveat with parallel chargers: There may be cases where a battery cannot safely accept the combined initial charge current of both chargers after a power failure. This might happen if the load on the dc bus is light when ac power is restored. In this case, you may have to reduce the current limit settings of both chargers.

What happens with a shorted cell?   1.5.5.3

Shorts happen. Through manufacturing defects, physical damage, or normal aging, a cell in a string may become shorted.[5] This reduces the back EMF of the battery by 2 Volts or so. In equalize charging, that 2 Volts must be absorbed by the remaining cells.

---

5   Shorts, of course, are relative. In flooded cells, if collected sediment at the bottom of the jar contacts the plates, it can form a high-impedance path for current which will cause increased self-discharge. We think of this as a "low quality" short, as opposed to a true short circuit caused by a fault in the connected dc equipment.

Example: In a string of 60 calcium-alloy cells, one cell is shorted. The battery is equalized at 2.25 VPC, for a total of 135 Vdc. The actual voltage on the 59 good cells is $135 \div 59 = 2.28$ VPC. While a difference of only about 2%, over an equalize period of 24 hours this difference could impose several excess ampere hours on the battery, and require more frequent water addition.

HOW DOES TEMPERATURE AFFECT BATTERY CHARGING?   1.5.6

We previously talked about temperature effects on batteries during discharge (Section 1.4.6). It's not surprising that temperature excursions can also affect batteries during charging. Only it's worse.

A lead-acid battery has a negative coefficient of on-charge voltage; as battery temperature increases, the terminal voltage required to maintain the same charge current decreases.

Example: A 125 Vdc battery floating at 132 Vdc at 77 °F (25 °C) might have a float current of 20 mA. If the battery temperature increases to 95 °F (35 °C), the charge voltage needs to be lowered to 128.7 Vdc to maintain the same 20 mA of float current.

Why should we care? The most important answer is that if we don't adjust the charging voltage at higher temperatures, we'll use a lot more water, and generate more hydrogen and oxygen, that must be vented out of the battery room. If the battery is an antimony alloy, it's possible to produce stibine, which, you remember, is poisonous. The gassing produces more grid corrosion. If it's calcium alloy, we may alter the grid's grain structure, reducing the amount of active material available for discharge. At higher temperatures, the self-discharge rate increases, and even keeping the float current the same may allow more sulfate buildup.

Basically, we'll wear out the battery even faster. There is already a shorter life expectancy for a battery that spends a significant amount of time at elevated temperatures – a reduction of 50% for each 10 °C (18 °F) of elevated temperature.

On the cool side, of course, we would have to raise the float voltage to maintain the same current, and a temperature-compensated charger will do just that. Without compensation, the battery might not reach full charge at low temperatures.

See more on *Temperature Effects* in Chapter 5.

DOES A "MAINTENANCE FREE" BATTERY NEED MAINTENANCE?   1.5.7

A battery is like your car: maximizing its useful service life requires routine maintenance. Even batteries advertised as "maintenance free" aren't. Yes, maintenance is reduced because you don't have to, or can't, add water to the electrolyte, and you can't measure the specific gravity. But you still need to keep the batteries clean, measure cell voltages on a regular basis, keep the intercell connections tight, and check for container cracks or potential leaks. If you find a leak, the offending unit must be replaced. Clean up any spills with sodium bicarbonate (watch for excess foaming); ammonia can also be used. Remember that lead-acid batteries are treated as hazardous materials, as is the electrolyte. Lead-acid batteries should be recycled (in fact, that's the law in most states).

The most important maintenance task that you need to perform for a flooded battery is to maintain the proper level of electrolyte. Here's how.

When and how do I add water?   1.5.7.1

Battery jars for flooded cells are transparent or translucent.[6] At the top of each cell section are two index marks, indicating the maximum and minimum electrolyte levels. The minimum level, as you would expect, is a comfortable distance above the tops of the plates.

Keep the electrolyte level roughly centered between the two marks. During discharge, as sulfuric acid combines with the active material of the plates, the level decreases because some of the sulfuric acid is combining with the active material in the plates. Likewise, the plates expand because lead sulfate takes up more room than the original active materials.

It's tempting to add water at the end of the discharge, especially if the level has fallen below the lower index. Don't. When you recharge the battery, the sulfuric acid returns to the electrolyte solution, and the electrolyte level in the cell rises above the index mark again.

The time to add water is at or near the end of a complete recharge. If the level is not midway between the index marks, add enough water to bring it to that level. If you're doing an equalize charge, you can add water during the last hour or so of the recharge, so that the gassing action during equalization will mix the electrolyte, avoiding stratification.

If a battery hasn't been maintained properly for a while, the tops of the plates may be uncovered. In this case, add enough water to cover the tops of the plates, and do a complete recharge. Then add water, as before, to get to the right level.

---

6 A photograph of a flooded cell with its index marks is shown in Figure 1c in SECTION 1.2.2.2.

» *Q: If I'm only adding enough water to get to the midpoint, why is there even an upper index mark?*

A: *The smart aleck answer is so that you'll know where the midpoint is. The real answer is so that you'll know if too much water has been added or added too soon in a charge cycle.*

Remember that the electrolyte level and specific gravity are related. When the battery is fully charged, and the level is correct, specific gravity will be at the manufacturer's specified value. If the water level is too high in a cell (that is, above the upper index mark), the gravity will be low, even if the cell is fully charged. This means that the cell voltage will be low. If the electrolyte level is high enough to leak through the vent caps, it's a good idea to remove some of the electrolyte. You will have to monitor and adjust the specific gravity over the next couple of charge cycles. If specific gravity is too low, battery capacity will be reduced. If it's too high, corrosion will be increased.

Does the amount of water required depend on battery design?   1.5.7.2

Water usage varies widely among the various alloys and grid designs used for lead-acid cells. The oldest grid alloy in use, a so-called lead high-antimony alloy, has the lowest potential for electrolysis (meaning it starts to gas at a lower voltage). It also has a relatively high self-discharge rate. For this reason, this alloy is used only for flooded cells, and in most cases only for pasted-plate cells. The antimony alloy is superior for cycling applications, such as motive power batteries.

To reduce the maintenance requirements of cells, manufacturers have developed several other alloys. Lead-calcium alloys have been in use since the 1930s. They have the opposite set of characteristics from antimony alloys: much lower float current and water usage, low self-discharge rate, but relatively poor cycling ability. They are ideal for standby and stationary applications and are the alloy of choice for VRLA batteries.

Low-antimony alloys, often containing selenium, have lower water usage than high-antimony alloys, while maintaining good cycling ability.

Lead-tin alloys, sometimes marketed as "pure lead" batteries, can be used for VRLA batteries in high-rate applications. You probably won't see this alloy in a flooded construction.

BATTERY CHARGERS

POWER CONVERSION: CHARGERS ARE POWER CONVERTERS?

WHY DC POWER? 2.1.1

Modern institutions rely increasingly on electrical and electronic controls. Where continuous service is critical, redundant power sources are needed to ensure reliability. Where locally-provided redundant power is required, the use of dc powered controls, operating on storage batteries, is one of the most reliable solutions. In stationary applications, ac power from the mains is converted directly to dc to provide power to controls and operating equipment, while maintaining a battery to provide emergency power. In motive power applications, the dc power is used only to restore the charge to the battery, which is used elsewhere.

Figure 2a: Typical charger used for stationary & emergency power applications

Most of the battery-powered installations in the US are based on lead-acid batteries. While the conversion of ac power to dc power is independent of the storage medium, the major focus of this section will be on chargers for lead-acid batteries, such as the one pictured in Figure 2a.

RECTIFICATION: HOW DO WE CHANGE AC TO DC?   2.1.2

In the beginning, there was the dynamo. Actually, there wasn't even a dynamo. Although Planté invented a practical storage battery in 1859, early experimenters had no way to charge it other than using primary batteries. Doing that today, we would quickly consume the world supply of zinc.

Anyway, back to the dynamo. Siemens and Wheatstone each developed practical dc generators, or dynamos, by 1867. This led to the first practical way to charge the infant lead-acid battery. When Camille Alphonse Faure invented the pasted-plate lead-acid battery in 1881, we finally had an easily manufactured source of stored dc power, and the battery could be easily recharged using a dynamo (usually steam powered). However, there was no ac power yet. No grid or mains. No substations. No transformers. These words weren't even in the English language yet. Also, no power failures.

But I digress. You want to know about changing ac power to dc power. Tom Edison was distributing dc power in limited areas in the late 1800s. You know how that turned out. By the mid-1890s, Westinghouse had won the "current wars," and was building widespread ac distribution systems. Now that customers had a source of ac power, they needed a means to convert it to dc to charge their storage batteries. That was the dynamo, mechanically coupled to an ac motor (the M-G set, which puts the ac motor part and the dc generator part in the same machine, came later).

Dynamos[1], of course, have some disadvantages, chief among them being weight, floor space, and maintenance requirements. By the 1920s, vacuum-tube or gas-filled rectifiers (ignitron, thyratron) were replacing dynamos in static converters (that is, no moving parts). Finally, by the 1960s, silicon diodes and SCRs (silicon controlled rectifiers), used in various circuit configurations, came to dominance in converter technology.

The rectifier circuit, of course, is at the heart of any battery charger. Although chargers are frequently called "rectifiers" in stationary applications, we'll use the term "rectifier" here to mean only the part of a battery charger that changes ac to dc.

---

1   There's more on dynamos and static converters in *Ripple* (CHAPTER 3) and *Output Current Limit* (CHAPTER 4).

POWER SUPPLIES: HOW RELATED TO BATTERY CHARGERS?   2.1.3

A battery charger, as you will see, is a special case of a dc power supply. Common technologies used in power supplies are discussed here.

Linear Power Supplies   2.1.3.1

Linear power supplies usually consist of a transformer, rectifier circuit, output filter, and control circuit. Since the rectifier operates at line frequency (60 Hz in the Americas), extensive filtering is required. They are capable of very accurate dc voltage control and are suitable for low power applications; they're frequently used for variable output voltages, such as laboratory power supplies. Their efficiency is limited unless designed for a specific constant load.

Ferroresonant & Magnetic Amplifier Designs   2.1.3.2

For high power applications (up to 100 kW or more), power supplies using magnetic controls have been developed. Ferroresonant supplies (aka "ferros") depend on the characteristics of a special transformer design, whose secondary voltage is fiercely independent of changes in the input voltage; that is, it's an excellent line regulator. The transformer has inherent current limiting, which by design could be as low as 200% of the power supply rated output. The rectifier circuit uses only silicon diodes, so the supply is extremely rugged. Efficiency is good at full load, and input power factor is high. A variant, the controlled ferroresonant transformer, improves the output voltage regulation at the expense of added complexity, and provides the capability to have separate float and equalize voltages. But, the transformer is larger than a standard linear transformer, generates more acoustic noise, and costs more. It operates only on single-phase ac sources; a three-phase power supply requires two or three transformers.

A magnetic amplifier (usually called "mag amp") power supply also uses a special transformer design, or a standard transformer coupled with another magnetic component, a saturable reactor. Like the ferroresonant supply, it uses silicon diodes in the rectifier, and is extremely rugged, although it lacks the high input power-factor of the ferro. The mag amp supply is available in single-phase or three-phase versions. One limitation of the mag amp rectifier is that the dc output current cannot be reduced to zero, so loading or bleed resistors are required for effective float-voltage control.

Phase Control Power Supplies   2.1.3.3

We've previously mentioned ignitrons, thyratrons, and SCRs. These are all examples of controlled rectifiers – that is, they are electronic devices that, like diodes, conduct current in only one direction. Unlike diodes, which have only two terminals (an input and an output), they have (at least) a third control terminal. They will conduct, or "turn on," only when given a signal to do so at the control terminal. This gives rise to phase-controlled rectifiers: the rectifier element (SCRs in modern supplies) conducts for only part of each cycle of the ac input power. By adjusting the conduction duration in each cycle, the power supply can control its output current from zero to its maximum rating.

The combination of a linear transformer and silicon controlled rectifiers makes for a very rugged supply. Phase control supplies can have very high efficiency, even at light loads, and have lower costs than mag amp or ferro supplies of the same rating. Because the output voltage waveform is discontinuous, however, the dc output requires more filtering, and there are more harmonics in the input current than for a ferro. But three-phase versions are simple and reliable, and don't suffer from potential instabilities that may occur in magnetically controlled supplies.

An exhaustive explanation of how phase control works appears in Chapter 3, *Ripple*.

Switching-Mode Power Supplies (SMPS)   2.1.3.4

All the power supply designs that we've discussed use transformers to isolate the dc output circuit from the ac power source, for safety reasons. The transformer also scales the output, by suitable selection of the secondary voltage, to the range needed for the dc voltage. So far, all the transformers have operated at line frequency, which means 60 Hz in the western hemisphere.

Line frequency transformers are large, heavy, and expensive. They're also unbelievably reliable. OK, so a pole (utility distribution) transformer in your neighborhood failed last summer, and you were in the dark for four hours. I assure you that's a rare event, compared to the number of transformers in service.

Switching-mode, or high frequency, power supplies have been in development since the 1960s. The switching-mode power supply (or SMPS) incorporates three power conversion stages: an ac to dc converter (rectifier) to provide a dc supply to an inverter, which drives a high frequency transformer to achieve isolation between the ac line and the dc output. Finally, another rectifier on the transformer secondary side produces the dc power for the load.

This seems a long way around the barn to get from ac to dc. The impetus, of course, is that the high frequency transformer is smaller, lighter, and less expensive than its line frequency counterpart, operating at frequencies from 20 kHz to several hundred kHz. The savings in the transformer can more than offset the extra cost of two more conversion stages.

Despite the multiple conversions, switching-mode chargers can reach efficiencies of at least 90%. Modern circuit design techniques can give them very high power factors. The switching components in the inverter stages, usually MOSFETs (metal-oxide-semiconductor field-effect transistors) or IGBTs (insulated-gate bipolar transistors), while not as rugged as SCRs, are achieving high reliability with careful circuit design, and can handle high power levels.

SMPSs are cost effective in power ranges up to many hundreds of watts for low voltage applications. They're the supply of choice for desktop computers because of their small footprint. Note, though, that the small footprint is obtained by including a cooling fan – not desirable in a remote application, such as a substation. The SMPS hasn't made significant inroads into higher voltage charger applications (125 to 480 Vdc), or stationary applications in general. Yet.

BATTERY CHARGERS: HOW DIFFERENT THAN POWER SUPPLIES?   2.1.4

There are just two qualities of a dc power supply that we need to control: dc output voltage and dc output current. Real-world dc loads have resistance, measured in ohms, and may be associated with a voltage source, such as a battery. So, to a power supply, a battery looks like an ohmic load and another power supply at the same time.

If we connect a resistor to the output of a dc power supply, we can calculate or measure the current; see the boxout *Ohm's Law* below. But if we connect a battery, the battery voltage opposes the power supply voltage. This is known as a *back EMF*. The current that flows is proportional to the difference between the power supply voltage and the battery voltage, following Ohm's law.

So, in the example in the following boxout, if the load is a battery with a voltage of 120 Vdc and an internal resistance of 1.25 ohms, and the charger is set to 125 Vdc, the current will now be 4 amperes (125 – 120, or 5 volts, divided by 1.25 ohms), instead of the expected 100 A.

EXAMPLE

By the way, 1.25 ohms would be an unacceptably high internal resistance for a secondary battery. I hope your battery never gets that bad.

It's the nature, then, of real world loads to respond to a voltage or current stimulus according to their ohmic value. That means that if you apply a voltage, the load decides how much current to draw. And if, instead, you apply a current, the load decides what the voltage will be.

KEY CONCEPT

OHM'S LAW

One way of stating Ohm's law for dc circuits allows us to calculate the current that will flow in a resistor, if we know the voltage:

I = V ÷ R

where V is the power supply output voltage, R is the value of the resistor in ohms, and I is the resulting current in amperes.

Example: For a voltage of 125 Vdc, and a resistor of 1.25 ohms, the current is 100 A.

If you know any two of the quantities in a dc circuit, you can calculate the third.

An Important Law   2.1.4.1

That leads us to the First Law of Battery Chargers: In a dc system with a battery, *the battery, not the charger*, determines the dc bus voltage. In the example above, the battery voltage is 120 Vdc, even though the charger is set to 125 Vdc, because 120 Vdc is the voltage the battery will maintain while receiving a current of 1.25 Adc. It appears that the battery voltage is 125 Vdc, but some of that voltage is lost in the internal resistance.

KEY CONCEPT

But, you say, when you set up your system, you're setting the float voltage. Right, but what you're really setting is the upper voltage limit for your power supply. A battery charger, you see, is a current source, not a voltage source. Your world is upside down.

If you have a regulated, constant-voltage power supply and a resistive load, you of course set the voltage where you need it. In most cases, you also set a maximum current that the supply can deliver, just in case your resistive load starts to demand too much current. But from zero current up to that current limit, your power supply provides the same voltage. This is how a current-limited, constant voltage power supply works. With only a resistive load, a battery charger looks just like a power supply, and works the same

KEY CONCEPT

as any constant voltage power supply.

KEY CONCEPT

But as soon as you connect a battery to the charger, things change. A battery charger, when connected to a discharged battery, becomes a current source. The current value is the current limit setting of the charger. From zero voltage all the way up to the voltage limit (float voltage), your charger provides the same current. This is a voltage-limited, constant-current power supply. When the battery becomes fully charged, things look more normal.

If this seems confusing, don't worry. You can still operate your charger normally, with all the settings you're used to, and everything will be fine.

UNINTERRUPTIBLE POWER SYSTEMS: USE OF INVERTERS & CHARGERS 2.1.5

In SECTION 2.1.3.4, which discussed SMPS, we used the word *inverter* without much of an explanation, other than that it is some sort of power conversion thing.

By now, you have a pretty good idea about the available methods for converting ac power to dc power. An inverter covers the other direction: changing dc into ac. Normally, when we speak of an inverter, we mean a device that generates line frequency power from a dc source, to power conventional ac-powered loads such as a personal computer, or industrial controls and instrumentation. A UPS integrates an inverter and battery charger in a single system, providing backup ac power for critical loads. Many modern UPS also include the battery and charging controls in the enclosure.

An inverter is different from an alternator, or ac generator, in that it's static (no moving parts), being based entirely on semiconductor circuitry. In a traditional inverter, a transistor (or SCR) circuit switches the current (from a dc source) in the primary winding of a linear transformer to generate an ac voltage in the secondary. The transformer is designed to provide the required ac voltage; it also provides the necessary safety isolation from the battery or other dc source. Using this method, the transformer operates at line frequency, which means that it's large and heavy, albeit reliable.

Inverters using line frequency transformers may have square wave or sine wave outputs. Square wave inverters are lower in cost, but are suitable only for simple linear loads, such as incandescent lighting. The square waveform has high harmonic distortion and may cause an unacceptable increase in temperature in transformer-operated equipment, such as another power supply or fluorescent lighting. They are also not very good at handling low power factor loads.

Line frequency inverters with sine wave outputs usually use ferroresonant transformers or brute force filtering and are generally more expensive than square wave inverters. An inverter described as "quasi sine wave" (or similar weasel-wording) is a square wave inverter with a limited pulse width, usually 120 to 126 degrees (see *Enter phase control* in Section 3.2.1.3 for some insight on pulse width). This output is a little easier to filter, but a little harder on the inverter transistors.

Pulse width modulation (PWM) is a technique that can be used to drive the primary of a standard transformer with a "chopped" square wave that results in a pretty good synthesized sine wave. This is the preferred technology in most high-power inverters and UPS. The advantage is that the output can be easily filtered to produce a good low-distortion sine wave. The disadvantage is higher complexity in the inverter circuit. Despite the complexity, the cost is a little lower than a ferro inverter. PWM inverters have been used in economical line-interactive UPS and in "on-line" or double-conversion UPS.

Modern inverter circuits also can use multiple power conversion stages, similar in operation to the SMPS, to generate a low-distortion ac output. The difference is that the last power conversion stage switches the isolated dc power to generate a line frequency ac, usually as a synthesized low-distortion sine wave.

DC-DC CONVERTERS: A SUITABLE ALTERNATIVE
TO BATTERY CHARGERS?   2.1.6

What if you had all the power you wanted or needed at 125 Vdc, but your site manager suddenly handed you a requirement for 24 Vdc to run some communication gear? You might tell him to buy a 24 V battery and a 24 V battery charger. But, these elements come with costs: initial, maintenance, and probable future battery replacement.

There is another way: for approximately the same cost as the battery charger, you can install a dc-dc converter. This product accepts a power input at 125 Vdc, and through internal magic, produces an output of 24 Vdc for your communications gear. You can eliminate the cost and maintenance requirements of another battery. Another advantage is that the output voltage won't vary: it's 24 Vdc regardless of whether the battery is on float or equalize, or the dc bus voltage is decreasing due to a power emergency.

Is there a downside? Well, of course. If the primary 125 Vdc battery fails, you lose communications. With a separate 24 V system, you at least would be able to call someone to inform him that the 125 V battery had failed. But, statistically, the converter approach is more reliable, since there is one less failure point.

A dc-dc converter is based on SMPS technology. Most designs are isolated, meaning that the output power is ohmically isolated from the input. A 125 Vdc system is usually floating; that is, the battery has no connection to earth ground, whereas many 24 Vdc systems are grounded (usually at the positive terminal of the battery). An isolated converter makes this possible; you ground the positive output of the dc-dc converter, and the floating operation of the 125V system won't be affected.

Isolation, in the sense we use here, is usually expressed in volts. The minimum isolation requirement for a 125 Vdc bus would be 1,250 volts, as specified in nearly every industry standard for power conversion equipment.

GENERATORS: WHAT ROLE DURING POWER EMERGENCIES?   2.1.7

Well, back to moving parts for a moment. There is another way to get emergency or standby power, at least of the ac variety, and that is with a gasoline- or diesel-powered generator. Most large generators are diesel powered; gasoline power is generally limited to small generators, such as for residences. Some large generators may use natural gas turbines. We'll discuss diesel generators, as used in emergency power applications.

Generators may be used stand-alone for emergency power, which implies that there will be an interruption in power availability between the instant of mains power failure and when the generator is started and up to speed. This isn't a viable situation for continuous process control, such as that needed in utility applications. In generating stations, backup generators may be used to provide emergency power to the charger/rectifiers and other ac loads during long-term power outages, where the battery backup time is insufficient. The generators may also be used to "cold start" remote stations, such as natural gas-powered turbines. In these cases, the generators are sized with spare capacity of at least 25%, in order to provide inrush currents for chargers, ac motors, and similar equipment.

In applications where a generator will be used for continuous duty, such as peak load shaving, it will be sized for the expected load with a small capacity margin, to avoid causing diesel motor damage from operating at a continuous light load.

CHARGER SIZING: WHAT CAPACITY DO I NEED?

HOW MUCH DOES EXTRA CHARGER CAPACITY COST?   2.2.1

It's no secret that you pay for power, and since power is voltage times current, you are going to pay more for higher voltage and for higher current. What isn't so clear is just what the differences are.

As the current output requirement of a battery charger increases, the cost of the components to handle that current increases proportionally to the current. That is, it costs about twice as much to provide a charger output of 100 A than an output of 50 A. This ratio isn't reflected in charger prices, though, because it doesn't take into consideration the fixed costs, such as control circuitry and cabinetry. When all that is included, the ratio is closer to the square root of the current.

Increasing the voltage has a similar impact. If you double the output voltage (but keep the same current), the cost of the power-handling components increases by about 1.4, but the fixed costs remain the same. So, it makes sense not to buy more charger than you need. There are intangible reasons, also. A larger charger might come in a larger enclosure, and you would need more floor space. You may need to beef up floor load-bearing capacity. You may also require additional ventilation. Plus, there's the challenge of getting it past your purchasing manager.

Well, you wouldn't buy more voltage than you need, at any rate, since the battery voltage determines the charger output voltage. But it's tempting to specify a higher current rating than you need initially, on the theory that the final system may somehow work better. It makes sense to overspecify if you're confident that the site's power needs will expand. But how do you know how much margin to order when the power requirements and the battery size are fixed and known?

Fear not. Help is right around the corner (or at least in the next paragraph). There is an easy method to calculate the charger size you need.

HOW DO I CALCULATE THE CHARGER CAPACITY I NEED? 2.2.2

In a true standby application, such as some substations, emergency lighting and alarms, and even engine starting, you can safely ignore the standing load when sizing the charger. The purpose of the charger is to charge the battery within a certain time limit. But, you still need to know how much the battery is discharged, which varies widely depending on the application. A safe approach is to assume that the battery is fully discharged at the point where ac power is restored to the charger. Then, the charger rating is Ah ÷ T, where Ah is the ampere hour rating of the battery, and T is the allowed recharge time.

Wait, there's more. Since the charge acceptance decreases near the end of charge, you need to add a little fudge factor to the Ah rating. Use 10% for lead-acid, and 40% for nickel-cadmium. So the complete formula becomes

$$A_{DC} = \left(\frac{AH}{T} \times K \right)$$

where A_{DC} is the charger ampere rating, and K is 1.1 or 1.4 for lead-acid or NiCd, respectively.

If you're in a generating station, or have another application with a constant load on the dc bus, you must add that load to the charger rating. If the load is variable, and you don't know the average dc current, then a safe approach is to use the maximum load current in your calculation (you can cheat on this if you know that the maximum is a very short-term load, occurring infrequently). Then the sizing formula becomes

$$A_{DC} = \left(\frac{AH}{T} \times K \right) + I_L$$

where I_L is the continuous load in amperes.

Here's an example: At a site with a 100 Ah lead-acid battery, a standing switchgear load of 21 amperes, and an eight-hour recharge requirement, you would rate the charger this way:

$$A_{DC} = (100 \div 8) \times 1.1 + 21$$

A_{DC} = 34.75 amperes. So you would order a 35 A battery charger. Of course, if you expect your site to grow, you might have to obtain a larger charger initially to allow for increased loads later.

» *Q: Isn't that cutting it close? There's essentially no margin there.*

   A: Unlike batteries, chargers don't decrease in capacity as they age. You don't need to add margin to compensate for age or the normal operating temperature range, as you do for a battery. The necessary margin is built into the calculation.

OUTPUT VOLTAGE RANGE: THE LONG & SHORT OF BATTERY STRINGS

IN SECTION 1.4.5, *How Do I Size a Stationary Battery*, the first step is selecting the number of cells you need, based on several factors: the acceptable voltage range for your dc-powered equipment, the float and equalize voltages for the battery type, and the end-of-discharge voltage, which is dependent on the discharge rate.

The number of cells normally falls into a small range. For example, for 125 Vdc buses, using lead-acid batteries, most applications are satisfied with between 58 and 62 cells, with the most common being 60 cells (Figure 2b). For nickel-cadmium, the nominal number is 100 cells, but is usually lower, around 92 cells, because of the high charge voltage requirement of the NiCd cell.

Battery chargers are designed with output voltage ranges that accommodate the usual

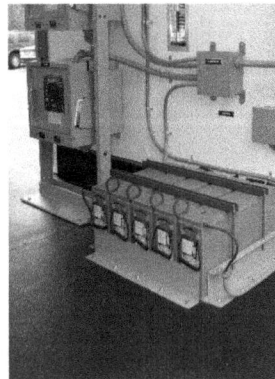

Figure 2b: Five VRLA battery containers on one wall of a trailer. Each has six cells, for a total of 30 cells. The other side of the trailer has 30 more. Combined it is a 60 cell, 120 V string, the most common combination

range of cell combinations. For a 125 Vdc bus, for example, a typical equalize voltage range extends to an upper limit of 147 Vdc, adequate for 62 lead-acid cells or 100 NiCd cells. Some NiCd batteries may require higher voltages. Check with the manufacturer. There are similar ranges at the other standard bus voltages to accommodate varying cell combinations.

The NEMA (National Electrical Manufacturers Association) standard PE 5 for stationary chargers specifies the float and equalize voltage ranges for lead-acid and NiCd batteries. Most available chargers meet or exceed the required ranges.

» *Q: I was just told that we need a new charger for an old, I mean old, emergency lighting system. It runs on 36 Vdc, using 18 old lead-acid cells. You don't have anything like that in your catalog. Where do I go?*

A: Not to worry. Application engineers at the manufacturer's factory can whip you up a custom charger in no time at all. All you need to tell them is the number of cells and the maximum equalize voltage. Oh, and of course, the output current.

AC POWER INPUT: WHAT FACTORS SHOULD I CONSIDER?

NOW THAT YOU know the charger rating you need, consider how to provide it with ac power. You can order a charger to operate on virtually any of the worldwide standardized ac mains voltages (up to a maximum of 600 Vac), as noted in Table 2a.

Table 2a: Worldwide ac mains voltages used to power chargers of various ratings for both single- & three-phase systems

Standard 60 Hz Voltages		Standard 50 Hz Voltages
Single Phase	Three Phase	Single & Three Phase
120 V	208/240 V	220 V
208/240 V	480 V	380/416 V
480 V	550 V (Covers 525-600 V)	
550 V (Covers 525-600 V)		

HOW MIGHT THE INPUT VOLTAGE SELECTED AFFECT COST?   2.4.1

The standard LVAC (low voltage ac) distribution voltages for the western hemisphere and for Europe are shown in Table 2a in the last section. Usually, utilities maintain a voltage within ±5%, but as you know, brownouts happen. NEMA PE 5 requires a charger to operate with an input voltage 12% below the nominal. This means that a charger wired for 120 Vac input can operate with an input as low as 105 Vac. At the high end, the NEMA standard allows the input voltage to be 10% over nominal.

As you know, the ac input current depends on your choice of input voltage: the higher the voltage, the lower the current. This could have an impact on your wiring costs. Always choose the highest branch ac voltage available at your site that is convenient for wiring.

Single-phase chargers can accept any voltage from 120 V to 480 V or higher. However, 120 Vac isn't available on some large charger ratings, such as 130 V at 50 Adc; for these ratings, you would select 208 Vac or higher. And of course, 120 V isn't available on three-phase chargers at all.[2]

You also may have a choice between single-phase and three-phase ac inputs. In general, a single-phase charger costs less than a three-phase charger, and the site wiring costs may be affected. Here's a little table (Table 2b) to help you decide, based on the popular 130 Vdc, 50 A charger rating:

Charger Model	AC Current @ 208 Vac	AC Current @ 480 Vac	Relative Cost
130 V, 50 A, Single Phase	72 A	31 A	1.00
130 V, 50 A, Three Phase	37 A	16 A	1.26

Table 2b: Relative cost when comparing single-phase & three-phase ac inputs for the popular 130 Vdc, 50 A configuration

Although you would need larger wire and a larger branch circuit breaker (see boxout *About that breaker…*) for the single-phase charger, you're running only two wires instead of three, and you have the offset of the lower charger cost. Of course, there may be other considerations, such as maintaining balance on a distribution transformer, that will dictate your use of a three-phase charger. Decisions, decisions…

---

2   Manufacturers sometimes see a request for "120 V Three phase." The user is probably distributing 208 Vac 4-wire power in the building, and is measuring the voltage from line to neutral. The charger doesn't care whether the supply is 3-wire or 4-wire, so you should order the 208 V input.

ABOUT THAT BREAKER...

Circuit breaker ratings can be complex. AC circuit breakers with trip ratings up to 100A use a hydraulic/magnetic trip mechanism. This breaker type is sensitive to the average value of the input current.

AC breakers with trip ratings over 100A use a thermal/magnetic trip mechanism, which is sensitive to the rms value of the input current.

What is the difference? The rms value is a mathematical representation of the effective, or heating, value of an ac current. Normally, the rms value of an ac input current will be a little larger than the average value.

Both circuit breaker types provide safe and adequate protection for the charger and the ac power source it's connected to.

An important specification for circuit breakers is the ampere interrupting rating. For more on this, see Standards & Codes in Chapter 9.

DETERMINING THE AC INPUT CURRENT   2.4.2

The ac input current is generally published in the charger manufacturer's literature. The published value is usually the maximum current under all worst-case conditions: extremes of ac input voltage, maximum output current limit, and maximum battery equalize voltage.

Chances are good that you'll never see an input current that high. First, you probably aren't equalizing the battery at the maximum available voltage, and the ac input current is usually close to the nominal value. But since the charger's ac input circuit breaker must be sized for the maximum possible current, your wiring and your branch breaker must do the same.

TECH TIP

This brings up a very important point. To meet agency certification requirements, the charger's input breaker must be rated to carry 25% more than the actual maximum input current (and could be as much as 100% higher). Since circuit breakers are available only in certain increments, you should rate your wiring and branch breaker according to the charger's input breaker rating, not the published input current rating.

Another important rule: You must size your branch circuit wiring according to the trip rating of *your* branch circuit breaker, not the input breaker in the battery charger. Your branch circuit breaker will be rated the same as, or larger, than the charger's input breaker.

There are some infrequent applications where you might order a charger with multiple input voltage capability, such as 120/240 Vac. The charger's input breaker trip rating has to carry the input current for 120 Vac, but you might actually wire the charger for a 240 Vac input. In cases like this, the ac input breaker might be rated for even more than 200% of the actual input current.

AC INRUSH CURRENT   2.4.3

Most available stationary chargers are based on circuits using line frequency transformers. Like ac motors, transformers have an inrush current when ac voltage is first applied, due to the initial magnetizing current requirement. The inrush current is large but lasts only for one or two cycles of the line frequency.

Maximum inrush is specified as a peak current, expressed as a multiple of the normal full-load ac current. If the full-load input current is 16 Aac (from Table 2b, Section 2.4.1), and the inrush is 15 times, then the peak inrush current is $15 \times 16 \times \sqrt{2}$, or about 340 amperes. The branch circuit breaker must be rated to withstand this inrush for at least two cycles; we recommend a medium or long time-delay circuit breaker. If your branch circuit breaker has an adjustable magnetic trip, set it to its maximum value. See the boxout on the previous page for additional technical details on circuit breaker ratings.

SOFT START   2.4.4

Soft start (also known as "walk-in") is unrelated to the ac inrush current. You're going to have an inrush current whether the charger is equipped with soft start or not.

The soft start feature slowly ramps up the dc output current from zero to its maximum required value, over a period of a few seconds to 10 seconds or more. It was initially added to battery chargers to prevent oscillation or overshoot in the output voltage when an underdamped charger[3] was connected to a relatively small battery. The requirement now appears in most charger specifications. Soft start also reduces transients fed back to the ac source, in case the charger is operating from a generator, since a generator may have a higher source impedance than a normal ac supply.

INPUT VOLTAGE & CURRENT HARMONICS   2.4.5

The ac voltage delivered by the utility to your battery charger, or any other device or equipment, is a pure sine wave, or at least it's supposed to be. If you would like more information on ac voltage and sine waves, see *Under the Hood: Sources of Ripple* in Section 3.2.

The current that is drawn by ac-operated equipment, though, might not be a pure sine wave. If the ac load is a pure resistance, such as incandescent lighting, the current will be sinusoidal. This is also true of loads like small ac motors. These are examples of *linear* loads; in this sense, linear means that the current follows the voltage, and isn't interrupted during any part of the ac line cycle. A linear load won't cause any harmonic currents or voltages to be generated; that is, its harmonic distortion is zero.

If there are linear loads, then there must be non-linear loads, right? A non-linear load is characterized by a load current waveform that doesn't look like the voltage waveform; that is, it's non-sinusoidal. It causes harmonic distortion. A phase-controlled battery charger is an example of a non-linear load.

You may be asking, "What's a harmonic?" A harmonic is an electrical signal that can be impressed on the ac line voltage or current that is some multiple of the base line frequency. We call the base frequency (60 Hz, for example) the *fundamental*, or *first harmonic*. Without harmonic distortion, this is the only signal present. Harmonic distortion introduces *higher order* harmonics: the second harmonic is 120 Hz, the third 180 Hz, and so forth. An example of ac line voltage containing multiple harmonics is shown in Figure 2c.

Non-linear, single-phase loads usually generate harmonics starting with the third, but only if the current has the same waveform for both the positive and negative voltages. Three-phase loads, such as a three-phase charger, generate harmonics starting with the fifth.

---

3   An underdamped control circuit achieves fast response but allows some voltage overshoot or undershoot. Conversely, an overdamped control circuit minimizes overshoot or undershoot, but at the expense of slower response.

Figure 2c: Measured ac voltage that exhibits multiple harmonics of the main 60 Hz sine wave

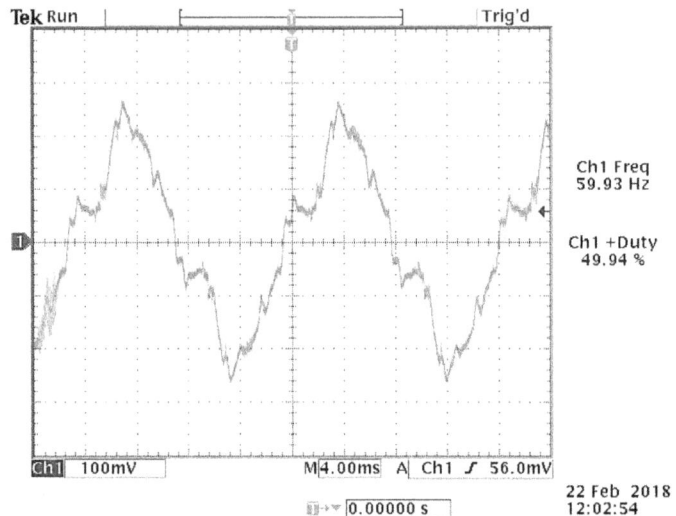

What's wrong with harmonics? They don't do any work in the battery charger or any other useful equipment (except for incandescent lighting or resistive heating). Instead they cause extra heating in the load, in the wiring to the load, in the distribution transformers, and all the way back to the generating station. Harmonic currents even force the utility to oversize the overhead distribution cables.

For any non-linear equipment, such as a battery charger, the input current may have harmonic distortion. The current distortion causes some fluctuation in the sinusoidal voltage that powers the charger; thus current harmonic distortion is translated to voltage harmonic distortion, which is then spread to other equipment in the facility. If the ac supply has a low impedance, such as that supplied by a standard distribution transformer, the voltage harmonic distortion that results is usually very low. Low harmonic voltage distortion is important to prevent excess heating in other loads connected to the facility's ac supply.

Power Factor in Linear Loads   *2.4.5.1*

We mentioned *power factor* in the description of various ac to dc conversion methods back in SECTION 2.1.3. You may be familiar with power factor correction, where a facility will install a capacitor bank to raise the power factor presented to the utility source. This is a good way to compensate for inductive loads, such as large ac motors or ballasted (e.g. fluorescent) lighting.

If the current in a load is linear, and exactly follows the voltage, the power factor is 1.0. This is the most desirable situation. A lot of loads, however, such as ac motors, are inductive. This means that the current has the same shape as the voltage but lags the voltage by some phase angle. Mathematically, we define power factor for this type of load as the cosine of the phase angle between the voltage and the current.

Examine the waveforms in Figure 2d. This is the relationship you might expect to see in a fractional horsepower motor. The applied voltage (trace A-A) is 120 Vac (rms), and the motor current (trace B-B) is about 14 Aac (rms). Note, though, that the peak value

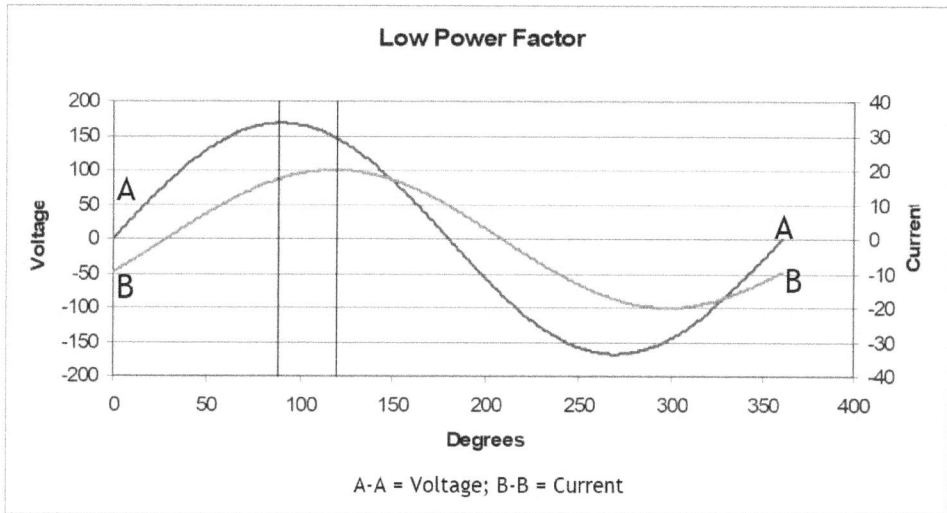

Low Power Factor

A-A = Voltage; B-B = Current

Figure 2d: Example of current lag in inductive circuit; the peak voltage at the first vertical occurs about 30 degrees ahead of the peak current. This phase difference results in a lowering of real power by a factor

of the current occurs later than the peak value of the voltage. We say that the current lags the voltage, which is what happens if the load is inductive. AC motors are inductive, especially if lightly loaded.

If you multiply the voltage by the current, you get $120 \times 14 = 1680$ VA (VA stands for volt-amperes). This isn't all real power, though. The real power is lower, reduced by the power factor. The motor produces real power only when the instantaneous voltage and current are positive, or when they're both negative. When one is positive and the other negative, it actually subtracts from the work done by the motor. By the way, power factor defined this way is also known as *displacement power factor*.

Vertical lines are drawn through the peaks of the voltage and current waveforms above. We can measure the time or phase difference between the peaks, in degrees. In the example above, it's about 30 degrees. The power factor, then, for the motor in this picture is the cosine of 30°, or 0.866. So the real power, with rounded off values, is:

120 Vac × 14 Aac × 0.87 = 1,462 watts

I hedged a little bit when I called this example a "fractional horsepower" motor. At 1,462 watts, it's a little over one horsepower (746 watts). To factor in average motor efficiency, you can use an approximate value of 1,000 watts per horsepower.

You see that with an inductive load, we can calculate power factor by measuring the phase angle between the voltage and current. But utilities don't do it that way. They say:

$$PF = \frac{watts}{VA}$$

where PF is power factor, watts are the real power in the load, and VA is the product of the voltage and the current in the load[4]. The VA product, or *volt-amperes*, is never less than the power in watts, so that you cannot have a power factor greater than 1.0. A power factor of 1.0 is ideal, since the entire current contributes to powering the load.

Power Factor in Non-Linear Loads   2.4.5.2

When you calculate power factor as the ratio of watts to VA, you find that non-linear loads also have a low power factor, even if there is no phase angle between the voltage and current, during the periods when current is flowing. Consider a simple power supply, with a simple diode rectifier, and a large filter capacitor connected to the output of the rectifier. The diodes conduct current only when the applied ac voltage is higher than the dc output voltage of the power supply, but there isn't a lag between the voltage and the current. Yet the power supply has a low power factor. Why? The answer is tied up in that definition for power factor, **W ÷ VA.**

The result is those pesky harmonics. When the current flows through the diodes for only part of a cycle, lots of harmonics are generated in the input current. But remember that harmonics don't do any work; they just add to the VA measurement at the input to the power supply. So, the power factor looks low – actually *is* low – because of the non-linearity of the load. This is the way a phase-controlled battery charger behaves, and why the power factor for a charger is less than 1.0, especially at light load.

At very light loads, you might not care about the power factor. For a substation application, where the float current is near zero, the power consumed by the charger is

---

4   Also known as *apparent* power.

low – just enough to cover the charger's fixed losses, such as power for the control circuits and the exciting current for the main transformer. The power factor in this condition will be low. The power factor increases as the dc output current increases, until the charger reaches its maximum power factor at full load. This usually ranges from about 0.8 to 0.9.

Incidentally, while the capacitor bank mentioned a few paragraphs back can compensate for an inductive power factor from linear loads, it does nothing for the low power factor caused by a non-linear load.

WHAT ALTERNATE AC POWER SOURCES CAN A BATTERY CHARGER USE?   2.4.6

Generators   2.4.6.1

Sites are sometimes equipped with auxiliary diesel generators, either to extend the backup time in power emergencies, or to provide power for a "cold start" for a remote or isolated power generator. Most battery chargers will operate fine with a generator power source, but the generator must be able to provide the necessary inrush current (see SECTION 2.4.3) when the charger is energized.

Generators are generally sized with some excess capacity in applications where ac motors or transformer-operated equipment are involved. For a full discussion of generator sizing, see *Can I run a charger from an auxiliary generator?* in CHAPTER 7, *Applications*.

Multiple Grid Sources   2.4.6.2

Site planners sometimes seek a level of redundancy by having a second ac power source available for critical equipment. Of course, the battery charger is critical, right?

The ac power input for a charger can be switched between two (or more, but it gets complicated) ac sources, using suitably rated contactors. There will be a short interruption of power to the charger, since the switching contactors need to be in a break-before-make configuration. Because of the interruption, the charger will also trigger the soft start feature, so the interruption of dc output current may be several seconds. Keeping the battery connected to the dc bus is tactically important in this kind of installation.

» *Q: Can a charger be powered by a UPS?*

A: *Well. The main question here isn't "Will it work?," but "Why?"*

A UPS is normally used to power critical ac loads during a power failure. A charger is normally used to maintain the charge on a battery so that the battery can power critical dc loads during a power failure.

Using a UPS to power a charger is robbing Peter to pay Paul. Backup time for ac-powered loads will be reduced in order to maintain the battery, which is presumably sized to power dc loads throughout a utility power failure. And it would be ludicrous to use a battery to power a UPS to power a charger to charge the same battery.

If you insist, there isn't any reason that a properly coordinated system won't work. If the UPS is a ferroresonant type, there could be some instability caused by ac harmonic currents in the charger input current. Also, the charger would take longer to "cold start," since the UPS might not be able to provide the ac inrush current for the transformer.

PACKAGING: WHY DOES IT MATTER?

WHAT ENVIRONMENTAL EFFECTS CAN MY CHARGER ENDURE?   2.5.1

NEMA Enclosures   2.5.1.1

The standard cabinet style for a battery charger is a NEMA 1 enclosure. NEMA 1 enclosures, such as the one pictured in Figure 2e, are made entirely of sheet metal, with provision for wall mounting or floor mounting, depending on the charger rating. Optional brackets are available for wall-mounted chargers, and some small floor-mounted chargers, to facilitate rack mounting.

Access to virtually all internal components is through the front door. The standard door requires a tool (usually a screwdriver) to open but isn't lockable. Again, there is an option to provide a padlock hasp for the door.

The NEMA 1 design is vented, usually through vent hole patterns on the top, bottom, and one or more sides, which makes it non-waterproof. Falling or dripping water can enter

Figure 2e: Example of battery charger in NEMA 1 enclosure during production before wiring has been completed

the enclosure, so it isn't suitable for wet locations or exterior installations (although drip shields are available that are effective against vertically dripping water).

For those locations, there are NEMA 12 and NEMA 4 designs. NEMA 12 is designed to resist dripping and splashing water, and the ingress of dust and foreign objects, and may be provided with knockouts for external electrical connections. It isn't intended for outdoor installation.

NEMA 4 is designed for indoor or outdoor installation. It is drip- and dust-proof, and will withstand hose-directed water. If a charger requires forced convection cooling, the NEMA 4 is fitted with watertight ventilation ports and dust filters on the ports. NEMA 4X is a corrosion-resistant construction, such as stainless steel.

IP Enclosure System   2.5.1.2

The IEC IP enclosure designation system defines construction features similar to the NEMA styles, but there is no exact correlation. An approximate equivalence can be determined from Table 2c.

IEC EQUIVALENCES	
IEC 529	NEMA 250
IP00	1
IP20	1
IP21	1 (with drip shield)
IP22	12
IP30	1 (with bug screening)
IP31, IP32	12
IP33	4
IP41, IP42	12
IP43, IP44	4
IP5x	4

Table 2c: Approximate equivalence between IEC 529 & NEMA 250 enclosure requirements. See Table 2d for explanation of 2 digits following "IP"

The degree of protection afforded by an enclosure is designated by "IP" followed by two digits (IPxy), that identify the level of solid object and water protection respectively, as listed in Table 2d. For example, IP22 protects against the ingress of moderately sized solid objects (approx. ½") and dripping water at an angle.

The meanings of IPx5 and IPx6 raise an interesting question, by grading the force of water jets. NEMA 4 enclosures are designed to withstand hose-directed water, without distinguishing between water jets and powerful water jets. The test specified

Numeral	First Digit (Dust & Solid Object Protection)	Second Digit (Water Protection)
0	Not protected	Not protected
1	Protects against solid objects of 50 mm diameter or larger	Protects against vertically falling water drops
2	As above, but 12.5 mm diameter or greater	As above, but with enclosure at 15° angle
3	As above but 2.5 mm diameter or greater	Protected against spraying water
4	As above, but 1.0 mm diameter or greater	Protected against splashing water
5	Protected against dust: ingress not prevented, but shall not penetrate in a quantity to interfere with satisfactory operation or safety	Protected against water jets
6	Dust-tight: no ingress of dust	Protected against powerful water jets
7	Not defined	Protected against temporary immersion in water
8	Not defined	Protected against continuous immersion in water

Table 2d: IP system for 2-digit (IPxy) rating for protection offered by an enclosure. The first digit (x) rates dust & solid object protection; the second digit (y) rates water protection

by NEMA to qualify a NEMA 4 enclosure uses a water jet delivering 65 gallons per minute. Is that powerful enough?

Note that all the enclosures defined above are designed for non-hazardous locations. There are additional NEMA styles intended for hazardous locations. Likewise, immersion in water requires special NEMA enclosure styles.

While we're on that subject, if your charger is ever in a flood, consider replacing it. Even if the charger appears to operate following a flood, persistent damage from immersion will cause problems at some point.

Rock-n-Roll: When the Earth Moves   2.5.1.3

Many chargers are mechanically designed to withstand some level of seismic activity. In the US, the expected severity of an earthquake is classified by seismic zones; zone 4 is the most severe. If you live in California, you already know this.

For example, HindlePower's AT series single-phase chargers are fully qualified for seismic zone 4, for installation at any point in the building (the tests are more severe for the upper floors of a building). AT30, the three-phase version, is qualified to zone 4 for the two smaller enclosures. The SCR/SCRF product line has no seismic qualification.

What altitude & temperature limitations do chargers have? 2.5.1.4

The method for sizing the battery charger presented earlier (SECTION 2.2.2) works for normal elevations, sea level to 3,000 feet, and for the charger's normal specified temperature range, 0° to 50 °C (32° to 122 °F). Outside those ranges, some derating is required. In other words, you have to purchase a slightly larger charger than the sizing calculation would indicate.

You can rate a charger for ambient temperatures up to 70 °C (158 °F), and for elevations up to 10,000 ft (~3,000 m). See *Temperature & Altitude Derating* in Appendix B.

» *Q: What happens if we store a charger in a very hot warehouse – say 175 °F?*

   A: Not a good idea. This is the same as 80 °C. While most electronic components tolerate high temperatures without permanent effects, some components, such as LCD displays, might be damaged.

WHAT'S AVAILABLE TO HELP IN PLANNING FOR CHARGER INSTALLATION? 2.5.2

Site planning requires accurate information on equipment installation requirements, such as overall dimensions, mounting or installation considerations, weight or floor loading, and locations of conduit entrances or other wiring provisions. Standard installation drawings are usually available on manufacturers' web sites, and the price is right. You may also be able to obtain documents for overall schematic diagrams, wiring diagrams, internal component layouts, and so forth. If you are ordering equipment with custom features, you can order an optional custom drawing package that will tell you everything you need for a successful installation.

WILL A CHARGER BE TOO LOUD FOR MY ENVIRONMENT? 2.5.3

In most cases, the sound level produced by a battery charger isn't above 55dB(A), and is described as a low hum, similar to what is produced by a distribution transformer. This level is roughly equivalent to normal conversation in a quiet room. So, acoustic noise often is not a factor in deciding where to locate a charger.

If the charger shares a work location such as an office, keep in mind that the charger may produce up to 65 dB(A) sound level when charging at full tilt. That's more like a loud conversation. But also consider that the battery will probably be nearby, with the inevitable release of fumes and gases, however slight. And, some people are more sensitive to noise, and may find even the low hum of a charger to be objectionable. You might want to build a separate closet for the system.

In case you're wondering where db(A) comes from, here's a quick run-down.

SOUND LEVEL

Sound, or noise, is measured by a logarithmic scale called *Sound Pressure Level (SPL)*. In the 1930s, Fletcher and Munson, working for Bell Labs, developed "equal loudness" curves for human hearing. Human ears aren't as good at hearing very low frequency sounds, especially below 100 Hz, as they are at the mid-range of normal voice frequencies, about 500 Hz to 3,000 Hz. Fletcher and Munson established 1.0 kHz as the reference point. They found that, for relatively quiet sounds, the SPL at 100 Hz had to be 20 dB louder for humans to perceive the same loudness as at 1.0 kHZ. For louder sounds, the difference wasn't as great.

The resulting curves, calibrated in *phons,* gave rise to three filter characteristics for measuring *Sound Level*, which is the term for Sound Pressure Level adjusted for the hearing contours. The filters are named – you guessed it – A, B and C. Filter A is used for virtually all acoustic Sound Level measurements in industry. A level of 65 db(A) is Sound Level, in phons. The A curve matches the ear's response at 40 phons; the B and C curves are used for higher Sound Levels.

The curves have been revised over the years, but the A filter remains the same.

RIPPLE & FILTERING

INTRODUCING RIPPLE

DC OUTPUT RIPPLE is a common and important specification for stationary battery chargers. The term "ripple" usually refers to the ac voltage measured at the battery terminals, but it may also be measured at the charger output terminals, if the battery is disconnected from the dc bus for maintenance.

If you're unsure about the causes and effects of ripple, then start with "Ripple," below. A discussion of dc filters, for reducing output ripple, starts in Section 3.4.

RIPPLE

A Play in One Short Act

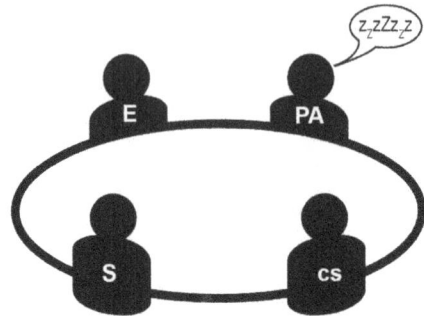

Dramatis Personae

E, *an engineer*

S, *a salesman*

PA, *a purchasing agent*

cs, *a customer service agent. He speaks in lower case because he doesn't know how to use the shift key.*

Scene

The lunchroom in a nondescript office building.

E is fumbling with a soda can, trying to open it. He's not sure which way to pull the tab. PA is seated in a chair in one corner, dozing. Every few minutes, he awakes with a snort, moves around a little, then falls back to sleep.

S enters. Seeing E, he immediately starts grilling him about a rumor that E is changing the design of the company's premier product, a stationary battery charger.

E: "Well, yes, we have something, but it isn't a change across the board. Call it an option."

S: "An option? What does it do?"

E: "We call it a filter. It takes all of the noise and ripple out of the dc output voltage."

S: "Wait a minute. Whoa up there. First of all, what is ripple? And what's wrong with our dc output that you need to screw with it?"

E: "Nothing's wrong with it, as far as it goes. But up to now, our only product offering has been a bare-bones charger, with no filtering. There's a lot of ripple in it."

S: "There you go again with that ripple. What is it?"

E: "Think of ripple as an unevenness, sort of a wiggle, in the dc output voltage that we deliver to the battery. Officially, it's a periodic variation in the dc output voltage."

S: "Periodic?"

E: "Yeah, that just means that it's consistent in its frequency content."

S: "I don't get that at all."

E: "Read Under the Hood: Sources of Ripple." [Section 3.2]

S: "OK, I'll do that later. But for now, what's the big deal with ripple? We get the battery charged, don't we? I never heard anyone complain about uneven voltage."

E: "You probably won't, as long as the battery is connected. The battery itself acts as a pretty good filter. But it's taking a bit of a beating doing it. And if the user ever disconnects the battery, say to change a cell, or just to clean the connections, all hell could break loose in the substation."

S: "Why is that?"

E: "With no battery, and no other filter, the ripple voltage could be high enough to damage equipment connected to the dc bus. That's especially true for a single-phase charger."

PA: (stirring slightly) "Ripple? Never drink it." (Goes back to sleep.)

S: (Ignoring PA) "So where does this ripple come from?"

E: "It's in the nature of how we create the dc voltage in the first place. A perfect charger would deliver pure, noise-free dc current to charge a battery."

S: "You mean just like the voltage we get from a battery?"

E: "Right. But the power for the charger comes from the ac power line, and has to be converted to dc. That's what the rectifier in the charger does. And you know that ac voltage is constantly changing, first positive, then negative, and on and on, 60 times a second." (E finally gets the soda open with a snap, spraying some of it on S's shirt.)

S: "Thanks, I needed that. Now, where does the ac voltage go, after you change it to dc?"

E: "That's the core of the problem. We have dc voltage, and can use it to charge a battery, but we also have a lot of the ac voltage left over, and that gets delivered to the battery, too. The ac voltage doesn't contribute to charging, but does raise the battery's temperature."

S: "I don't see that as a big deal."

E: "It isn't, for a large flooded battery with lots of excess electrolyte. But in a VRLA or gelled electrolyte battery, it can shorten the life of the battery. And too much internal heating could actually cause thermal runaway, which can destroy a battery."

S: "OK, that could be a big deal."

cs enters.
He had been standing in the doorway, listening.

cs: "hey, i just had a call yesterday from a guy at the fifth of may substation. he said they had to replace the whole battery because 3 of the units were all melted inside. is that what you're talking about?"

E: "Could be. But it probably just got too hot in the substation, or they had some other problem."

S: "I'm almost convinced. But how do you get rid of the ripple, and how much is it going to cost, and who pays? Us or the customer? I'm going to have a hard time selling this, if it increases the price."

cs: "c'mon, you know the answer to that last one. we always end up paying."

E: "Well, not necessarily. The filter option might add 15% to the cost of a charger. But in most places, the battery costs more than the charger, and extending its life will more than pay for the option. Customers who can see five years into the future won't have a problem with it."

PA is sitting up straight now.
He's been listening to the conversation for the last few minutes.

PA: "This filter option. What's in it? What do I have to buy now? And where are we going to store the stuff? The stockroom's full as it is. The boss keeps telling me to reduce inventory."

E: "Good point. Yeah, you'll need some space to stock more inductors and some big capacitors."

cs: "people are going to ask me what this filter thing actually does. can you give me some specs?"

E: "Thought you'd never ask. Take a 130 V charger in a substation. If the charger is unfiltered, the ripple voltage on the battery might be 1 or 2 volts. But take away the battery, and the ripple voltage could shoot up to a hundred volts. And the peak voltage, as I said before, could be high enough to damage the substation electronics." *(Note that E said 'as', not 'like'. My kinda guy.)*

cs: "so, how much ripple do you get with the filter in the charger?"

E: "We'll have two levels of filtering. With the standard filter, the ripple would be only 0.1 volt when the battery is connected, but might rise to a volt or so when the battery is disconnected."

S: "That sounds a lot better. You said you had two levels. What's the other?"

E: "We'll call it enhanced filtering. This would keep the ripple below 0.1 volt even if the battery is disconnected."

S: "You can call it anything you want, but we'll call it Battery Eliminator filtering. How much is that going to cost?"

E: "It'll add about another 10% to the charger. But customers that have really sensitive equipment, especially communications stuff, will want it."

cs: "a lot of customers won't want to pay extra for any filtering. how about if i just tell them to be sure to turn off the charger first, if they're going to disconnect the battery?"

E: "Sure, that'll work. You want to take the chance that they'll always do that?"

E finishes his soda and leaves the room.
S looks down at his shirt, and shakes his head slowly.

THE END

UNDER THE HOOD: SOURCES OF RIPPLE

BEFORE YOU GET started here, you'll find the material in Chapter 2 on battery charger technology helpful.

UNAVOIDABLE BYPRODUCT OF CHANGING AC POWER TO DC POWER.   3.2.1

Ever since Tesla and Westinghouse showed us the way to make ac power transmission practical, we've spent a lot of time and effort developing better ways to change ac power back into dc power.[1] We need dc, of course, to charge batteries, and also to power critical equipment that must operate in the absence of the primary ac power source. Following is an overview of the methods we have of converting ac power to dc.

What are some early ways of changing ac to dc?   3.2.1.1

In the beginning, we used rotating machines. Can you say maintenance? Size? Weight? Noise? Ozone? As soon as the mercury arc rectifier became commercially viable (in the 1920s), industries moved quickly to adopt static converters (that is, no moving parts), although small M-G (motor-generator) sets persisted until the 1960s.

Development of the ignitron, thyratron, and other controlled rectifiers enabled *phase-controlled* rectification or conversion, which provides a much easier path to controlling the output voltage and current of a converter. With the exception of small thyratrons, however, these devices had a large drawback: a large pool of mercury, essential to its operation.

Enter the *mag amp* (magnetic amplifier), an older technology dusted off and used extensively in WW2. In the 1960s, magnetic amplifier converters were developed that used transistors and semiconductor diodes to replace vacuum tubes and gas-filled or mercury rectifiers. These solid state converters are very rugged, and have low maintenance requirements, but require complex and expensive magnetic components.

---

1   AC power transmission is the dominant method for short grid connections. High voltage dc transmission is an increasingly active field, especially for very long distances, undersea cables, and other applications.

So chargers mostly use SCR now? What's that?   3.2.1.2

In the late 1950s, GE developed and commercialized the *SCR* (silicon controlled rectifier).[2] Like a diode, it conducts current in only one direction, so it's ideal for converting ac power to dc. But it's like a diode on steroids: We can force it to conduct for a full cycle, or for any part of the cycle we wish.

A cycle?   3.2.1.2.1

Right. At this point, we can't avoid a graphical explanation. You heard E, in the play earlier, explain that ac voltage changes from positive to negative and back, 60 times a second. If you could see ac voltage (and your eyes were fast enough), it would look like the waveform in Figure 3a.

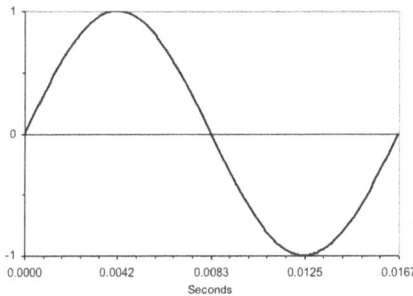

Figure 3a: 60 Hz ac voltage waveform

At the beginning of this picture of an ac waveform, the value of the voltage is zero. It rises to a maximum positive value, returns to zero, then goes to a peak negative value, and finally returns to zero. That is, it alternates; the positive and negative halves together make one full cycle. And it does all this in 0.01667 seconds (or 16.67 milliseconds)! Then it does it again. And again.[3,4]

This is a sine wave. There are other useful waveforms, such as square waves, but utilities like to generate sine waves because they require less energy than any other waveform to generate and waste less energy in transmission.

Now we need to change the ac voltage into a dc voltage. Since, by definition, dc voltage isn't supposed to alternate, we'd expect it to be always positive, or always negative. We'll play with positive. Here's a picture of a rectifier circuit (see Figure 3b), consisting of one diode, connected between an ac voltage source and a piece of dc-powered equipment.

2   Nowadays, it's more proper to say "semiconductor controlled rectifier," since some are fabricated with silicon carbide, for extra high temperature operation.

3   In North America, one cycle at 60 Hz = 16.67 msec per cycle. In most of the rest of the world, ac power is transmitted at 50 Hz, taking 20 msec per cycle. Hz (hertz) is the name given to frequency; one hertz is one cycle per second.

4   Note: All the waveform pictures in this section show exactly one cycle of ac voltage, unless noted otherwise. This section is concerned primarily with single-phase converters.

Ideally, we would like the voltage waveform we apply to the dc-powered load to be a smooth, non-varying straight line, just like we would get from a battery, unwavering for

Figure 3b: Half-wave rectifier (single diode)

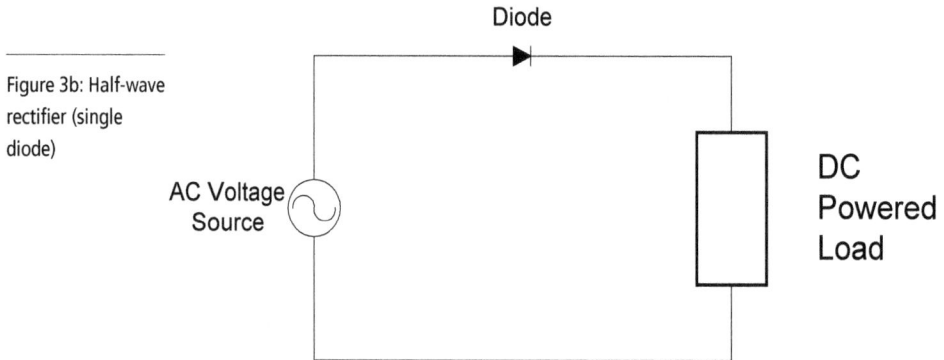

all time. Ain't gonna happen. Since the diode only conducts in one direction (conventional current flows clockwise in Figure 3b), only the positive half of the ac voltage waveform gets passed to the load as seen in Figure 3c:

Figure3c: Output waveform for half-wave rectifier

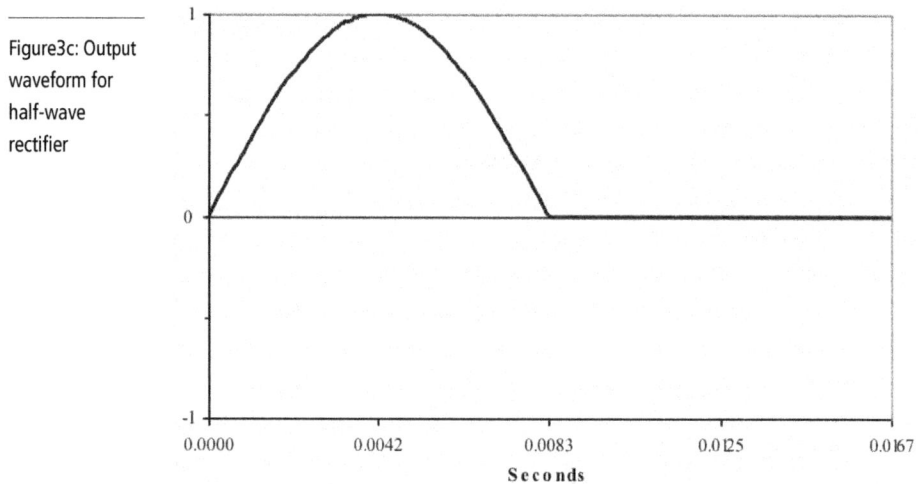

As you can see, the voltage applied to the load is always either zero or a positive value, and current will flow to the load only during one half of the cycle. So it's a dc value, but it has a lot of ac left in it. There are rigorous mathematical methods to determine just how much and what kind of ac voltage is left, but we're going to spare you. Instead, we'll take the next step, in order to make the ac to dc conversion a little more efficient. Instead of a single diode, we'll use a *diode bridge,* so that current flows during both halves of the ac voltage cycle (see Figure 3d). This is called a *full-wave* rectifier; the preceding circuit, you might have guessed, is a *half-wave* rectifier.

Figure 3d: Full-
wave rectifier
(using 4 diodes)

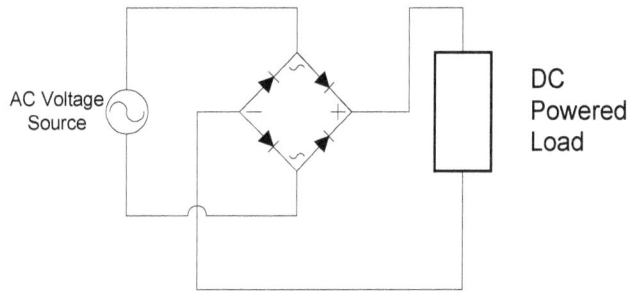

AC Voltage
Source

DC
Powered
Load

I see this new circuit uses four diodes.
Can't we do it with two diodes?   3.2.1.2.2

Actually, you could, but it makes the main transformer in the charger a little more complicated and expensive. Diodes, even with their heat sinks, cost less than the more expensive transformer. The downside is that each diode consumes a little energy, so this configuration generates a little more heat than the two-diode version.

Anyway, the advantage to a full-wave rectifier is that the negative half cycle, which was simply blocked in the half-wave rectifier, gets inverted, and becomes another positive half cycle when it's delivered to the dc load. Figure 3e shows the dc output from a full-wave rectifier:

Now you can see that the voltage applied to the load is positive most of the time. Current flows into the dc load for the whole cycle, not just for half of it. It might not be readily apparent, but this voltage waveform has less ripple than the half-wave version. And, of course, it's primarily dc voltage, which powers the load.

That's a pretty messy waveform.
How much dc voltage is actually in there?   3.2.1.2.3

Glad you asked. Since this rectified wave originates from a sine wave, it obeys some pretty consistent laws. The *average dc value* (that's the part that powers the load) is 0.637 times the peak value of the voltage. So if we want to control the dc voltage applied to the

Figure 3e: Output
waveform for
full-wave rectifier

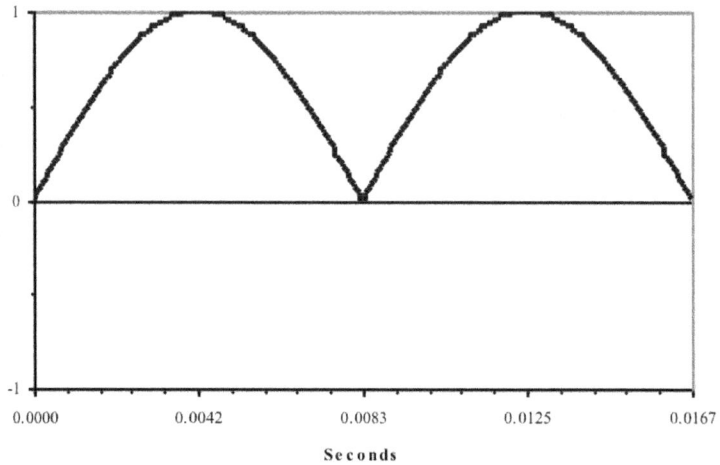

load, all we have to do is control the peak value of the ac voltage, which normally comes from a distribution transformer, or a transformer in the rectifier. Easy, eh?

No.

We wish it were that simple. But dc equipment (including battery) voltage requirements aren't always the same, the ac power line voltage isn't always the same, and if you're using a temperature-compensated charger, you know that the temperature varies, probably more than anything else. What we need is a way to control the average dc voltage output of the rectifier while everything around it is changing.

Before you take off on that, just how much
ripple is there in that full-wave rectifier?   3.2.1.2.4

KEY CONCEPT

In the full-wave rectifier? About 48%. That means that if you have a 125 Vdc rectifier, the ac ripple you're applying to the load is about 60 Vac. But remember a couple of things: we've been talking about a pure, relatively simple rectifier, connected to a resistive dc load – not a battery. Because a battery's terminal voltage will "push back" against the rectifier, the waveform changes, and your ripple results may vary. Also, remember that it's the peak value of the ripple voltage that may damage equipment.

Now, since you didn't ask, I'll tell you anyway. The lowest ripple frequency in the full-wave rectifier is 120 Hz, twice the frequency of the ac power line. But in the half-wave rectifier, it's 60 Hz, the same frequency as the ac power. 60 Hz ripple is much harder to

get rid of than 120 Hz ripple, another good reason to use a full-wave circuit, and stay away from half-wave rectifiers.

> I'm curious now. How much ripple is
> in the half-wave rectifier?   3.2.1.2.5

It's in the neighborhood of 121%. Sorry you asked? By the way, the ripple numbers here are the percentage of ac ripple voltage in the average dc voltage from the rectifier. That means that if you had a full-wave rectifier that gave you 1.0 Vdc (average) output, the ac ripple voltage would be 0.48 V. Notice that we give the ripple values as a percentage of the average dc value, not the peak value of the original ac waveform.

> Isn't there any kind of rectifier that produces
> less ripple, instead of more ripple?   3.2.1.2.6

Yes, and that's one of the big advantages of three-phase power. (Remember Nicola Tesla? He developed the concept of three-phase power transmission that's so widely used today. No, he didn't develop an electric car.) Even the simplest three-phase rectifier has only one-third the ripple of a single-phase full-wave rectifier, and it gets better from there. But remember that these numbers are for resistive loads.

> OK, now tell me how you're going to
> control that output voltage.   3.2.1.2.7

Thought you'd never ask. We mentioned earlier that several technologies for ac to dc conversion were developed over the years, and each had a unique method of control. For rotating machinery (motor-generator, or MG sets), you could vary the excitation field on the generator. You could also vary the speed, but since the generator is driven by an ac motor, that's harder to do. Mag amp and controlled ferroresonant circuits are easier to control; all you have to do is vary the control current in the winding(s) of one or more of the magnetic components. Each of these methods has its advantages. But we would like a method that doesn't depend on complex transformer and reactor design.

Enter Phase Control   3.2.1.3

Do you remember that, a few paragraphs ago, a famous writer mentioned the SCR? To refresh your memory, an SCR is like a diode on steroids: We can ask it to conduct for a half cycle, or any portion of a half cycle, in a process called phase control.

> First cycles... now you hit me with phase.
> Are these things related?   3.2.1.3.1

Couldn't have said it better myself. Remember Figure 3a that pictured a single cycle of ac voltage, starting at zero volts, and ending up at zero volts? Now imagine that waveform broken up into 360 little bits (degrees, actually), with the start at zero degrees and the end at 360 degrees. You may be saying, "Oh, just like a circle." Which is true, since a sine wave is mathematically related to a circle.

Now take just the first (positive) half of the sine wave. It runs from 0 to 180 degrees, and the peak voltage occurs at 90 degrees. You can think of each of these degrees as being a *phase angle*. Now consider that a regular diode, in a rectifier circuit, conducts current for the whole 180 degrees, the duration of the half cycle. If we could tell the diode to conduct for only half of the half cycle, well, we would have an SCR. When we command the SCR to start conducting current, we are *turning it on*. And if we could tell the SCR to conduct current for only half of the half cycle, through the magic of an SCR Gate Trigger Circuit, we would wind up with a voltage waveform sent to our dc load that looks like the bottom waveform in Figure 3f.

There are a couple of new features in this picture. Note that the phase angle, in degrees, is shown along the bottom, instead of seconds, although the time elapsed, in seconds, hasn't changed. Also note that the original ac waveform, which is the voltage applied to the input of the rectifier, is shown at the top; the output of the rectifier is shown under it. The vertical axes show the values of the waveforms. In this case each voltage waveform has a peak value of 1 volt.

Figure 3f: Original 60 Hz ac voltage input (top); bottom shows output from a half-wave SCR rectifier with phase control

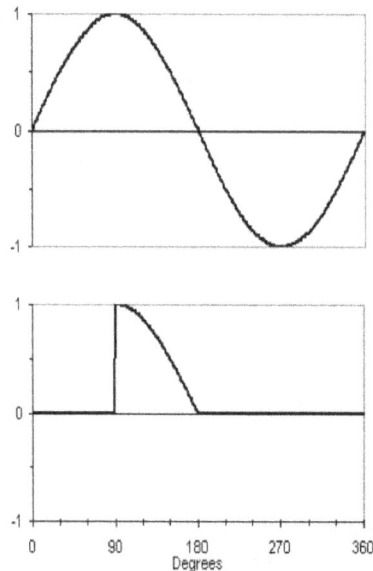

You can see that the voltage applied to the dc load is zero until the phase angle reaches 90 degrees. Then the SCR turns on, and the output of the rectifier then suddenly jumps to the peak value, which is the value of the ac input voltage at 90 degrees. If we have a resistive load, the current does the same thing. With a resistive load, the output voltage and output current have the same waveform, which is also the waveform of the incoming ac voltage.

But earlier you said that a battery isn't like
a resistive load. What happens in that case?   3.2.1.3.2

Good question. You might suspect from the previous picture that the ripple in the output of a controlled rectifier (that is, a circuit using one or more SCRs) is definitely going to be different from a half-wave or full-wave rectifier using standard diodes. However, if you connect the output of a controlled rectifier directly to a battery, the ripple might not be higher. Other components in the rectifier circuit can change the ripple values by modifying the output waveforms.

Before you start modifying waveforms,
tell me how I can make the SCR conduct
for the first 90 degrees instead of the
second 90 degrees.   3.2.1.3.3

Uh, I can't. It would be nice, of course, if we could make the SCR conduct for however long we wanted, anywhere within the cycle. But the way an SCR works is that it keeps conducting forever after we turn it on, as long as current is flowing through it. It turns off (stops conducting) when the current through it is reduced to nearly zero by other circuit components, or by the ac power source itself. If we wanted to turn it on at zero degrees, and then have it turn off at, say, 90 degrees, we would need some method of forcing the current to go to zero just when it's at its maximum value. We don't do that with ac to dc converters.

A type of SCR called a Gate Turnoff (GTO) SCR could be made to conduct for the beginning of a half cycle (that is, the first quarter of the cycle), and then be turned off for the second quarter. They're useful in inverters but are not normally used in ac to dc converters.

There are circuits that will do that (inverters, for example), but they're more complex and more expensive, and unnecessary in a rectifier. You can see in the preceding figure that the voltage goes to zero at 180 degrees; with a resistive load, the current also stops at that point. At this point, the SCR turns off by itself. It gets some help by the fact that the ac voltage on the SCR goes negative in the next half cycle, guaranteeing that the SCR current comes to a screeching halt. We call this *line commutation*, where line refers to the ac line, and commutation is the process of turning off the SCR. So, basically, if we want to control the voltage or the current in a rectifier (for example, by increasing the current from zero to a higher value), we start turning on the SCR at the end of the half cycle and move the turn-on point (phase angle) toward the beginning to increase the voltage or current. The point where we turn on the SCR is called the *firing angle*.

> I get that. I'm a commuter, too. But you
> only have a half wave there. Can you make
> a full-wave SCR circuit with phase control?   3.2.1.3.4

Now you're really getting it. Using a bridge circuit consisting of just two SCRs and two diodes, we can get power for the dc load that looks like the bottom waveform shown in Figure 3g.

Note that we get a pulse of output voltage in each half cycle. As before, this waveform will have less ripple than the half-wave version for any firing angle of the SCRs, but the half-wave and full-wave circuits both have more ripple than a simple rectifier using only diodes (for the same average dc voltage). Once again, for a resistive load, the current waveform in the load would look just like the voltage waveform.

> **I'll buy that. Now let's see what happens
> when you connect the rectifier to a battery.**   **3.2.1.4**

OK, you've been patient, waiting for this. Remember the picture of the waveform for a full-wave rectifier? Now consider this: if we connect a battery to the rectifier, the battery terminal voltage is going to push back against the output voltage of the rectifier. But since the battery voltage is constant, and the rectifier output isn't, the rectifier voltage is going to be greater than the battery voltage only part of the time. Take a look at this three-part waveform (Figure 3h).

> Wow! Now you're really getting complicated.   3.2.1.4.1

Yes, there's a lot more to chew on in this picture. Note the horizontal line cutting right through the rectifier output voltage pulses in the middle graph. That line is the dc voltage of the battery. Note that the rectifier output voltage pulses exceed the battery voltage only a small part of the time. Then, in the bottom graph, you see a waveform representing the current flowing from the rectifier into the battery. The current, as you see, flows only when the rectifier output pulse is larger than the battery voltage.

Figure 3g: Original 60 Hz ac voltage input (top); bottom shows output from a full-wave SCR rectifier with phase control (2 SCRs & 2 diodes)

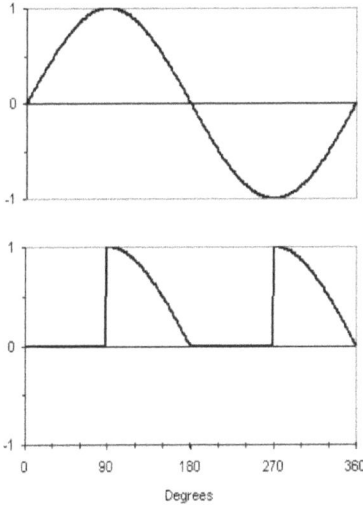

Figure 3h: Original 60 Hz ac voltage input (top); middle shows full-wave SCR rectifier output voltage & connected battery's dc voltage; bottom shows current from rectifier into battery

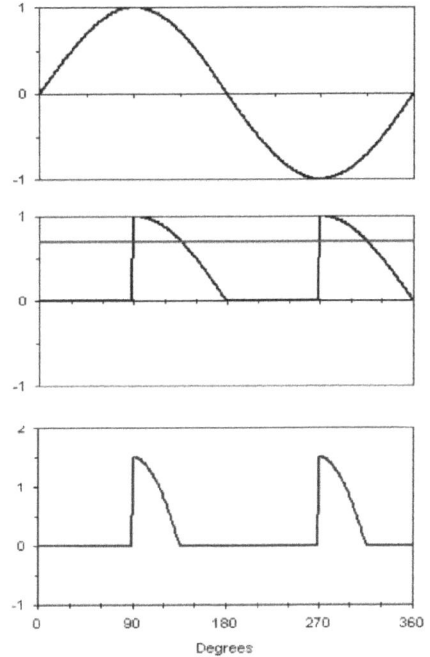

> Why doesn't the battery voltage change,
> and just follow the voltage of the rectifier?   3.2.1.4.2

That's a good question. In this example, we've kept it simple by assuming that the battery is a perfect voltage source. In real life, the battery is a very good constant voltage source, but you will see the voltage change a tiny bit when the current flows from the rectifier. And that tiny change is the source of the ripple voltage that appears on the battery terminals.

> OK. A couple of observations here. First,
> because the current pulse is so short, it looks
> like most of the voltage from the rectifier is wasted.
> Second, how do you know much current
> you're actually getting?   3.2.1.4.3

You're hired. You've just found the essential shortcomings of this simple rectifier circuit. Take the current magnitude: since there isn't any extra impedance between the rectifier and the battery, you could get a lot, *a lot*, of current flowing. As you know, current limiting is an essential feature of a charger. The only limit to current flow here is the inherent resistances of the wiring, the transformer, and the battery, which are all deliberately kept very low (these are known as *parasitic* resistances). We could put in a separate current-limiting resistor, but if you want to talk about wasted power....

And you recognized the other limitation: The current flows for only a brief time in each half cycle. This leads to higher ripple and more power losses. Fortunately, there's a single solution to both shortcomings, in the form of an inductor.

Enter the Inductor   3.2.1.5

An inductor is also known as a choke or coil. This is a magnetic part, separate from the main transformer in a charger. It acts as an ac impedance.

> An ac impedance? I thought you were
> trying to limit the dc current.   3.2.1.5.1

Oh, we are. This isn't magic, but it might seem that way. An inductor stores energy in a magnetic field, and we can use that feature to solve the aforementioned problems.

> I thought you wanted to charge a
> battery, and now you're playing around
> with storing energy?
>
>   3.2.1.5.2

KEY CONCEPT

Right. Here's the way it works. We put the inductor between the rectifier and the battery. It has just two terminals – an input and an output. Whenever the rectifier voltage is larger than the battery voltage, the inductor takes the energy caused by that voltage difference and stores part of it in its magnetic field. Then, when the rectifier voltage decreases below the battery voltage, the inductor uses that stored energy to continue to deliver current to the battery. In this way, we force the current to flow for a longer period than shown in the previous waveforms. Larger inductors–that is, inductors with more *inductance*–are more effective at smoothing the current. Obviously, there's a trade-off between inductor size and cost on one hand, and current smoothing on the other.

> Is that all there is to it?   3.2.1.5.3

Well, "all there is" has a lot of benefits. First, we reduce the ripple current, and that reduces the ripple voltage on the battery. Second, we reduce the *rms* current in the SCRs, transformer, etc., which makes life easier for them. Third, we get to use most or all of the voltage from the rectifier, instead of just a small part. I believe that was one of your objections earlier.

So, rms voltage multiplied by rms

It was. But now you've hit me with
another zinger. What's rms?   3.2.1.5.4

RMS (or rms) stands for root mean square, a mathematical description of dynamic waveforms. You don't have to remember that. The only thing you have to know is that it refers to what we call the *effective* value of an ac voltage or current. It's the measure of the heating value of ac power. We use it so we can equate dc and ac values. For example, a light bulb that takes 1 ampere at 120 V will consume 120 watts. If it's dc, we measure average voltage and current; if it's ac, we measure rms voltage and current. That 120 V, available at your standard wall outlet, is actually 120 V_{rms}.

So, rms voltage multiplied by rms
current will give me rms power, right?   3.2.1.5.5

Well, no. It's really average power. The term "rms power" has no real technical meaning. It's just a buzzword used by the audio geeks.

How does adding an inductor to the
rectifier change the waveforms?   3.2.1.5.6

That depends on just how much output current we need, and how large the inductor is. We try to design the charger so that the inductor is "critical" at the extremes of operation. That is, at the lowest battery voltage we would expect to see on a *normally discharged battery*, and the highest ac input line voltage, all at full rated output current. Since I know you're going to ask, I'll tell you that a critical inductor is one that is sized to force each SCR to conduct for a full half cycle, and that a normally discharged battery is one that has been discharged to 87.5% of its "nominal" open-circuit voltage, measured while load current is still flowing. For example, a 60-cell lead acid battery, with a nominal open-circuit voltage of 120 V, would be discharged to 105 V. That number may be different for some battery types and in some applications.

Finally, to answer your last question, here's an example of what the output current waveform might look like with a critical inductor, forcing the current in each SCR to be continuous for the entire half cycle (see Figure 3i).

This is an example for a 130 Vdc charger. Notice that the current has a peak value of about 60 A, but a minimum (valley) value of 40 A, and never goes to zero; thus, it flows continuously, even though the voltage waveform is discontinuous. In this example, the

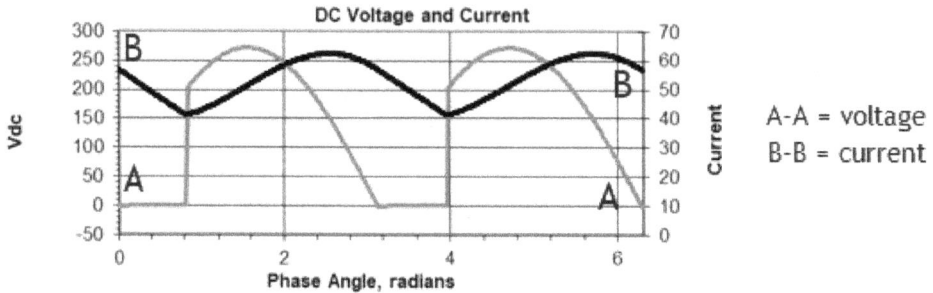

Figure 3i: Output from full-wave SCR rectifier with phase control combined with critical inductor, demonstrating current smoothing from inductor. Both output current & voltage are shown

SCRs are gated on at about 48 degrees. Don't let the phase angle in radians throw you; that's just an alternate to degrees (180 degrees is approximately 3.14, or pi, radians).

> I think I understand this. With continuous
> current flow, the ripple will be lower.   3.2.1.5.7

Right on the mark. The ripple voltage on a 200 Ah battery for the conditions shown above would probably be less than 30 mV (millivolts).

> But, but, but... the output current ranges from
> 40 A to 60 A. How can the ripple be so low?
>
> 3.2.1.5.8

Aha! That's through the magic of the effective capacitance of the battery working in tandem with the main inductor. You're right that the ripple *current* is high; notice I said that the ripple *voltage* was reduced at the battery terminals. And interestingly, the ripple *current* shown above is only about 7 Arms. To understand all that, continue reading to the section on filters.

WHY IS RIPPLE BAD FOR MY DC SYSTEM?

WE'VE MENTIONED, THROUGHOUT the previous sections, the bad stuff that too much ripple can do to your dc system. Here's a quick summary:

• High ripple currents can cause unnecessary battery heating, possibly shortening life.

- High ripple voltages can degrade the performance of equipment connected to the dc bus.
- If the battery is removed (or becomes defective), ripple voltage could be high enough to permanently damage connected equipment.

You've probably seen that we talk about both ripple voltage and ripple current. They're two sides of the same issue. In the next section, on filters, we'll describe the charger circuit components that we use to reduce both the ripple current and the ripple voltage that the charger delivers to the load and/or battery.

Before we go on, though, we should review the role the battery plays in reducing ripple voltage. Note that we're now looking at ripple voltage, not ripple current. The charger is basically a current source, and that includes the component of ripple current in the dc output. Thus, the charger delivers some amount of ripple current to the connected load. The resulting ripple voltage that you can measure on the output terminals depends on the nature of the load.

If the load is purely resistive (there aren't many, but incandescent lighting comes to mind), then the ripple voltage is directly proportional to the ripple current, obeying Ohm's law [$V_{RIPPLE} = I_{RIPPLE} \times R_{LOAD}$]. We've seen that in an unfiltered rectifier, ripple current can be pretty high, leading to high ripple voltage for a resistive load.

Most dc loads, though, have some inductance or capacitance. The granddaddy of capacitive loads is a storage battery; a healthy battery has lots and lots of effective capacitance. As you'll see below, that capacitance acts to reduce ripple voltage by 90% or more when compared to a purely resistive load.

FILTERS: HOW DO THEY REDUCE RIPPLE?

WHAT LEVELS OF FILTERING DO MANUFACTURERS OFFER?   3.4.1

Manufacturers handle filtering requirements differently, but most offer two levels of filtering: standard filtering and battery eliminator filtering. Both are adequate to reduce the ripple voltage at the battery to safe levels. Battery eliminator filtering offers additional filtering to ensure that connected equipment operates reliably even if the battery is disconnected for maintenance.

IS IT SAFE TO USE AN UNFILTERED CHARGER?   3.4.2

TECH TIP

We can't emphasize this enough: never buy an unfiltered charger. If you do, never disconnect the battery. If you do, never energize the charger.

Why? The output voltage of an unfiltered charger can have peak ripple voltages at least 50% higher than the nominal dc output voltage. Those peaks can damage sensitive equipment connected to the dc bus. Imagine plugging your desk lamp into a 240 Vac outlet. You'd have a lot of light for a little while. Maybe.

As we have seen, ripple, a small but regular variation in the dc output voltage and current of a charger, is an unavoidable consequence of converting ac power to dc power. Large amounts of ripple can affect battery life and pose a risk of equipment damage. Reducing the magnitude of ripple voltage and/or current is the job of a dc filter.

» *Q: Why do you call it a dc filter, if ripple is ac?*

   A: True, we're filtering out an ac ripple signal. Because it's on the dc side of the rectifier, though, and the output of the filter is almost pure dc, we call it a dc filter.

BASICS: A LOOK AT RIPPLE   3.4.3

Take another look at the current waveform in Figure 3i. You can see that the current contains a lot of ripple; that is, there's still a lot of ac in the output current, as well as the steady dc value of about 50 A. What you can't see in that figure is the ripple voltage on the output dc bus, because we don't yet know anything about the load: is it resistive, or is there a battery connected?

If you have ripple current, you're going to have ripple voltage. If you have ripple voltage, you're going to have ripple current. How much of each depends on the type of rectifier, the actual output current, the nature of the load, and... oh, yeah, whether we have a dc filter in the circuit.

MAIN INDUCTOR: A GOOD START; SMOOTHS OUT THE CURRENT 3.4.4

As we saw in the section on ripple, a simple full-wave rectifier (with a resistive load) can have as much as 48% ripple on the output voltage. In a resistive load, that would be 48% ripple current. But as you saw in Figure 3i, adding the main inductor, as pictured in Figure 3j below, reduced the ripple current to only (roughly) 7 Arms. The way the math works, this amounts to only about 1% ripple, a vast improvement over 48%.

You may be asking yourself (I hope you are) how 1% ripple could damage equipment. The answer is that this is a snapshot at one operating point, which is the full load for this charger. At very light loads, the inductor becomes less effective at reducing the peak output voltages.

ENTER THE CAPACITOR: SMOOTHS OUT THE VOLTAGE 3.4.5

An inductor isn't the only component with desirable ac characteristics. It does its work by impeding any tendency of the output current to change with changing voltage (hence, ac *impedance)*. What we would like to have is an element that resists changes to voltage as the current changes. Wait, wait... we do. A *capacitor* has a very useful characteristic: We can apply to it a varying current, in this case the 7 A ripple current, and it will have relatively small changes in the ac voltage on its terminals – much smaller than a resistor with the same current.

So, if we put an inductor in series with the output current, it will tend to smooth out the current. If we put a capacitor in parallel with the load (resistor or battery; it doesn't matter), it will tend to smooth out the voltage on the load.

KEY CONCEPT

Figure 3j: Schematic showing main inductor added to full-wave rectifier

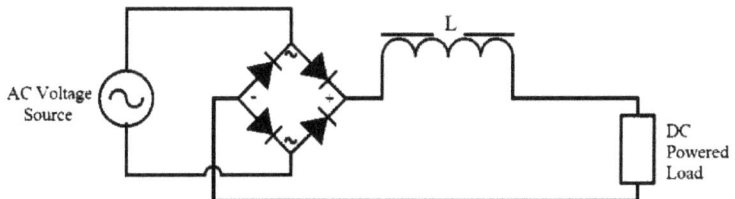

Now we can draw the original full-wave rectifier, with a filter added that includes a single inductor (L) and a single capacitor (C), pictured in Figure 3k.

Let's now look at some oscilloscope snapshots from a typical 130 volt, 75 ampere charger with an SCR rectifier that includes a filter consisting of a single inductor and single capacitor. These snapshots will help us see how these filter components smooth out the ripple in both the voltage and current delivered by the charger.

The following oscilloscope snapshot (Figure 3l) is the voltage before the inductor 'L'. This is the voltage, unfiltered, coming straight from the rectifier. The voltage clearly has strong ac characteristics and significant ripple.

The current flowing through the inductor appears in the waveform shown in Figure 3m. Note that this current is measured with a clamp-on current probe which explains why the vertical scale is in millivolts (mV) and not amperes. The scale translation is 1 mV to 2 amperes such that each major vertical grid equates to 20 amperes. While the current waveform clearly still exhibits ac ripple, as mentioned earlier when discussing Figure 3i, the inductor greatly reduces the amplitude of the ripple.

We can see that the presence of the capacitor in this filter smooths out the voltage delivered by the charger by comparing the last unfiltered voltage waveform (Figure 3l) with this next voltage waveform (Figure 3n). The waveform in Figure 3n is the voltage from the same charger across the filter capacitor (C) just after the inductor (L). The voltage now appears to be almost entirely dc, and the ripple voltage will typically be below 100mV when connected to a battery if the correct inductor and capacitor are selected.

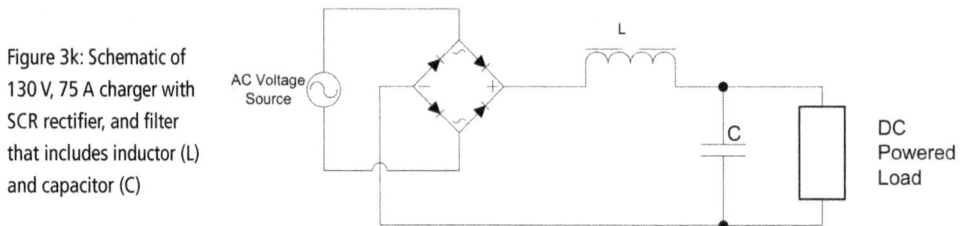

Figure 3k: Schematic of 130 V, 75 A charger with SCR rectifier, and filter that includes inductor (L) and capacitor (C)

While this filter, consisting of a single inductor and single capacitor as in Figure 3k, will work fine, we usually include a second inductor between the capacitor C and the load, forming a three-element filter. This improves the filtering performance when a battery is connected. A battery eliminator filter includes yet another capacitor between that second inductor and the connected load (you guessed it: a four-element filter); this ensures that you have the same filtering level in situations where there is no battery connected in the circuit.

Figure 3l: Unfiltered voltage
from rectifier pictured in
Figure 3k.
Voltage is measured by
an oscilloscope before the
filter's inductor (L)

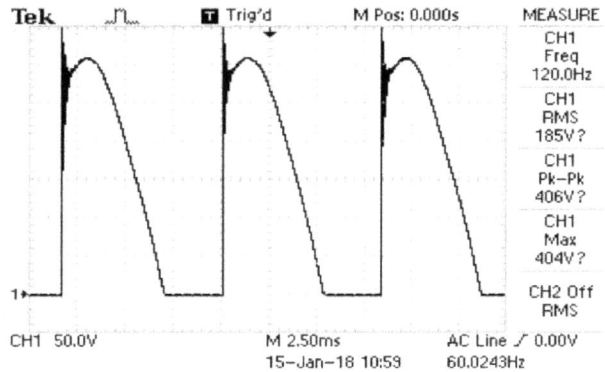

Figure 3m: Filtered current
from rectifier pictured in
Figure 3k. Current is
measured by an
oscilloscope at the filter's
inductor (L) with a
clamp-on current probe

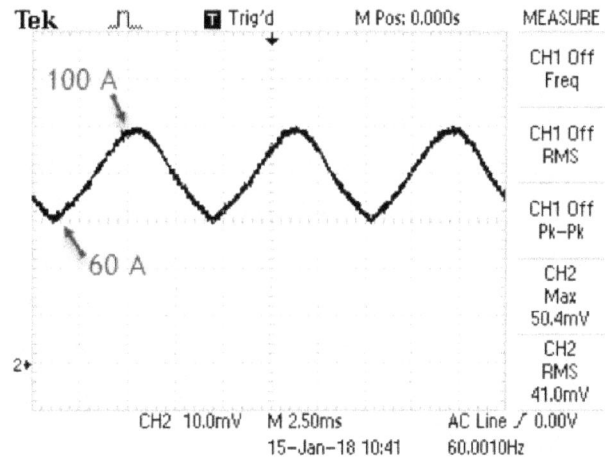

Figure 3n: Filtered voltage
from rectifier pictured in
Figure 3k.
Voltage is measured by an
oscilloscope across the filter
capacitor (C), just after the
filter inductor (L)

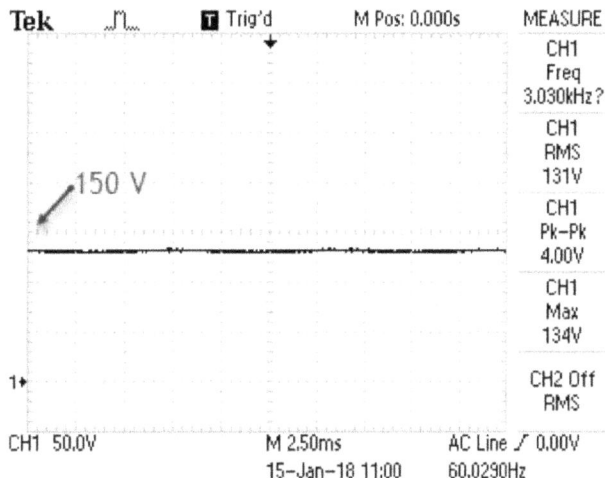

BEYOND RIPPLE: OTHER ELECTRICAL NOISE

DB OR NOT DB? THAT IS THE QUESTION   3.5.1

Now that you've read the preceding section on ripple, you know that ripple is measured in millivolts (mV). But suppose you're suddenly presented with a request to "limit the voice-band noise to 32 dBrnC." Where did this come from? You know that the charger makes a little sound, called *acoustic noise* (see SECTION 2.5.3), and that it's measured in dB(A). But this looks different. And what do they mean by voice band? Consider the following conversation.

DBRN? UNDERSTANDING C MESSAGE
WEIGHTING NOISE LIMITS   3.5.2

My customer is asking for the charger to have
32 dBrnC acoustic noise. What is this?   3.5.2.1

Actually, that spec conflates acoustic noise and electrical noise. dBrnC refers to electrical noise, not acoustic noise. The dBrnC number is a specification designed to limit the electrical noise that the charger produces in the voice band in telecommunications equipment. Think of the hum or static that you sometimes hear on the telephone.

What do you mean by voice band?
The charger has a dc output, right? Is the
charger going to start talking?   3.5.2.1.1

No, but this spec has to do with telecommunications equipment that may be powered by the dc bus. An example of this could be a telephone system. The customer wants to be sure that any noise on the dc output of the charger won't interfere with voice communications. An order for a charger or power supply will commonly specify 32dBrnC, but sometimes asks for the more stringent 26dBrnC.

He's ordering a filtered charger. There shouldn't be any noise.   3.5.2.2

Almost right. A charger's dc filter reduces the ripple, which is measured in millivolts (mV), and is usually 30 mV or less for a filtered charger. But the ripple frequency is 120 Hz for a single-phase charger. The voice band is higher, running from about 300 to about 3,500 Hz. The dc output voltage of the charger can, and usually does, have some noise in this frequency range. Your customer wants to be sure that the noise in those frequencies is reduced as much as possible. Usually, the ripple filter takes care of it. You can add the dBrnC requirement to a charger order, and ask the manufacturer to test for compliance, to be sure the charger is compatible with the telecom equipment.

So what does dBrnC mean?   3.5.2.3

dB (decibel) is a relative measurement of noise, including acoustic noise. That's why there is sometimes some confusion, because chargers also have an acoustic noise spec, measured in dB(A). But in this case we're measuring electrical noise.

"rn" simply stands for reference noise. The reference in this case is a power level of 1 picowatt (pW). That's a pretty low number – one millionth of a microwatt.

And the C refers to the filter used to shape the frequency response inside the voice band. The shape of the filter is shown in the figure below. When we use this filter to measure the noise, we say that we have a "C message weighted" signal.[5]

I looked at that curve (Figure 3o). You also have something in there called CCITT something, but the two curves look almost the same.   3.5.2.3.1

CCITT is the old name for the ITU (International Telecommunications Union) agency that prepares telecom standards. In other words, it's the international standard, equivalent to the C Message weighting standard used in the U.S.

---

5   C message weighting is different from the C filter response used for measuring acoustic noise. For more on this, see the boxout *Sound Level* in SECTION 2.5.3.

Let's see if I understand. There's
electrical noise on the charger output.
We'll pass that through a C filter, and
measure the voltage. But how do we
get that translated into dBrnC?   3.5.2.3.2

Well, we're using a special test set, of course. The test set has the filter built in, and it also provides a 600-ohm standard load. When we measure the filtered voltage on that 600-ohm resistor, we can calculate power. The test set is already scaled to a reference level of 1 pW, and so we end up with a measurement in dBrnC.

Or, we can just stop when we get the power, if we've been asked to meet the psophometric weighted limits.

The sophomoric?   3.5.2.3.3

The psophometric weighting curve is the one specified in the ITU-T standard 0.41. That used to be called the CCITT standard. You noticed earlier that it's almost the same as the C weighting curve. Customers sometimes ask for a psophometric measurement, but usually they want a C message weighted spec.

I think I follow. But I need to know the
bottom line. What am I supposed to order?   3.5.2.3.4

A charger with a standard dc filter will meet the 32 dBrnC requirement. To meet the 26 dBrnC spec, order a battery eliminator filter.

TECH TIP

Is that all there is to it? Why didn't you just tell me that
up front?

Figure 3o: C-Message & CCITT 0.41 weightings. Courtesy of Lindosland through public domain licensing. Found at English Wikipedia under "Psophometric Weighting"

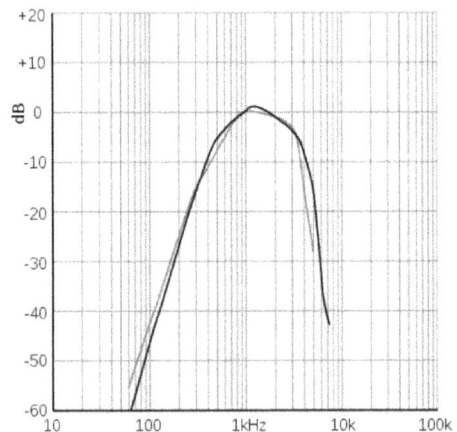

OUTPUT CURRENT LIMIT

THIS SECTION DISCUSSES why a battery charger may be equipped with a feature to limit the dc output current to the battery and/or load. Features for doing so from one manufacturer, HindlePower, are given for simplicity, but you can apply the principles discussed when considering similar options for any manufacturer's products.

WHY LIMIT CURRENT?

BEFORE WE DISCUSS what the dc output current limit feature does, let's start with what it doesn't do: it can't affect the ac inrush current on startup (see the discussion of inrush current in Section 2.4.3).

As the name implies, the dc output current limit feature controls the maximum dc output current that a charger can deliver to the battery and/or load. In most modern phase-controlled chargers, current limiting is electronically controlled. This control normally allows the user to adjust the current limit setting over a limited range. Typically, there is an upper limit to the allowable range built into the control circuit, to protect the system against risky maladjustment.

Why is there an upper limit? It isn't, as you might guess, to protect the battery, which is largely self-limiting as long as the charger output voltage is set properly.[1] But there are some practical reasons to limit the current:

1. To keep the output current within a safe range for the ratings of the charger components (rectifier diodes or SCRs, for example), and for the dc wiring connecting the battery to the charger.

---

1 Note however, that a large charger with a relatively small battery could be a mismatch, causing excessive gassing early in the charge (remember the ampere-hour law, Section 1.5.1). A stationary battery can be safely charged at an initial rate of about C/6. Chargers in parallel could have the same problem.

2. To control the ac input current to ensure that site wiring ampacities and circuit breaker or fuse ratings aren't exceeded.
3. To prevent nuisance trips for input or output circuit breakers in the event of a bus fault, or during unusual operating conditions such as starting a dc motor.

A LITTLE HISTORY

REMEMBER THE DYNAMO? No? Neither do I. Dynamos, or dc generators, usually driven by an ac motor, were among the earliest methods for recharging a storage battery. Originally, self-excited or shunt-wound dynamos were used because they had an inherent current limit if the dynamos became overloaded. The overload would reduce the output voltage, thus reducing the excitation, and the output current would "fold back" toward zero as the overload increased. The presence of a battery as a load, however, reduced the fold back, and effective current limiting required electromechanical regulators (similar to those in older vehicles with generators, before the switch to alternators).

The current regulators weren't user-adjustable; they ran at a fixed current, and tolerances were wide. Early dynamos typically limited their output current to about 120% of their rating, but could range from 100% to 125%. As electronic controls became available, the regulation specs improved.

When dynamos were replaced by static converters in the mid-1960s, and then finally by modern electronically controlled chargers, manufacturers perpetuated the 120% current limit, largely because it was in customers' specifications.

SOMETHING FOR NOTHING?

BUT WHAT DOES this limit really mean? In many users' minds, it has become a minimum value that the user could expect the charger to deliver. Remember that, originally, it was a guarantee that the current wouldn't exceed 120% of the rated value. You could still expect that the charger, or dynamo, would deliver 100% of its rating, and the current would be limited in all cases to somewhere between 100% and 120%.

Some users thought, of course, they could always depend on that 120% - that they were getting a 120 Adc charger for the price of a 100 Adc charger. Do you really think that you're going to get that for the price of a 100 Adc charger?

Consider the variations in the operating conditions for a battery charger. It's called upon to charge a battery that's been discharged anywhere from a few percent down to almost zero volts, up to its maximum permitted equalize voltage, for a stated maximum number of cells. And it must do that even with an ac power brownout (12% below nominal ac voltage) or up to 10% over the nominal voltage. Over a temperature range of 0 °C to 50 °C. With a connected load ranging from nothing up to 100% of its rating.

A combination of extreme operating conditions – ac brownout (-12% voltage), maximum equalize voltage for the highest possible number of cells, and a full load condition – defines the target for worst-case charger design. The charger is guaranteed to deliver 100% of its rated current under those conditions, but it isn't guaranteed to deliver more than 100%.

KEY CONCEPT

When a battery charger reaches current limit the dc bus voltage is determined only by the battery, not by the charger. When the ac power is restored after a failure, the charger will be in current limit, because the battery has been partially discharged, if only by a little, and its terminal voltage will be below the float voltage. Until the battery gets recharged to the float voltage, the charger remains in current limit. The charger output voltage will match the battery voltage. Period.

Typically, a battery charger doesn't run at the extreme conditions noted above, and so it may be able to provide more than 100% rated current. What we're saying is that you shouldn't count on it.

HOW DOES CURRENT LIMIT WORK WITH VARIOUS CHARGERS?

SCR/SCRF PRODUCTS   4.4.1

Current limit is specified to be adjustable from 80% to 120% of rated output. In practice, the adjustment range is wider; you can probably get a range of 50% to 125% or more. But please note that everything in the charger is designed for a maximum of 120%; trying to get more output current can overstress internal components.

Another important part of the current limit specification is the transient response. The question is, what happens when there's a sudden load demand, and the output current tries to shoot past the current limit setting? You could have this situation, for example, while trying to start a dc motor, which has a large inrush current. In HindlePower's SCR/SCRF charger, there is a short delay, about 100 ms, in the operation of the current limit circuit. During this time, the charger typically can deliver up to four times its rated output current. Because of this response, the charger contributes some of the motor inrush current; the rest is supplied by the battery. Obviously, if ac power fails, the battery must supply the entire inrush current.

Figure 4a shows the response you might expect to see during motor starting. This graph shows the result of a test at a gas turbine site. A 130 Vdc system includes a lube oil pump, rated at 60 A full load, and a three-phase SCRF charger rated at 75 A full load. This graph is for one combination of charger, battery, motor, and site arrangement, which includes how the dc wiring is run. Your results may differ.

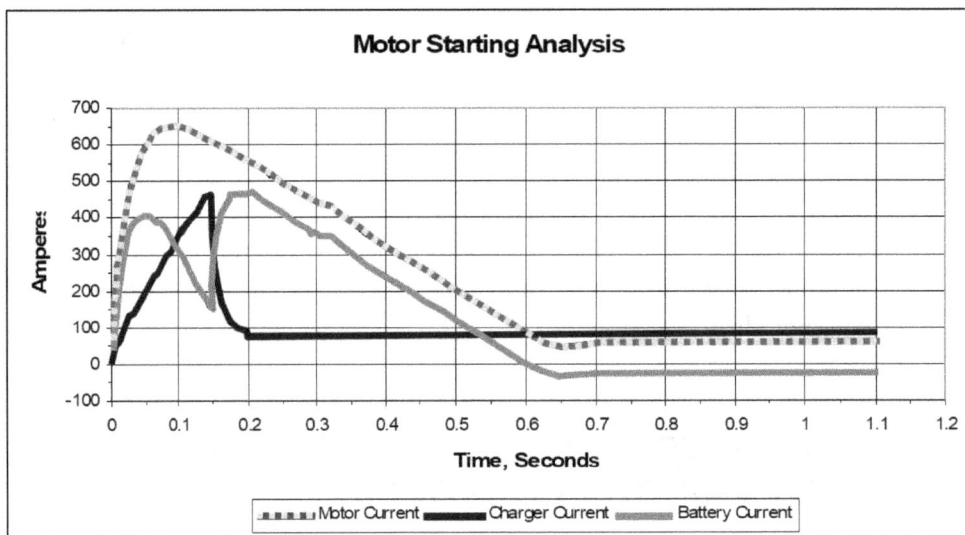

Figure 4a: Current response at startup of motor at gas turbine site. 130 Vdc system includes a lube oil pump, rated at 60 A full load, and a three-phase SCRF charger rated at 75 A full load

Note that the motor inrush current (dashed trace) reaches 640 A, about ten times the motor's rated full load current. This is typical. The inrush current is shared by the battery (medium gray trace) and charger (black trace). As the charger starts to limit the output current (at about 150 ms), the battery picks up more of the load, until it supplies the entire load current, less the 83 A provided by the charger (current limit was set to 82.5 A, which is 110% of the charger rating). When the motor is up to run-speed and its current drops to about 60 A, the charger supplies all the motor current, and also starts to recharge the battery.

At this site, an inrush-limiting motor starter wasn't used. A starter reduces the peak inrush current at the expense of startup time; instead of reaching full speed in less than a second, the motor takes several seconds to start, and starting torque is reduced. Incidentally, the charger at this site is equipped with a "motor starting" option, which replaces the standard dc fuse with a slightly slower rating, allowing it to survive the 150 ms delay in current limiting action.

Try to avoid a bus fault (a "bolted" short circuit on the dc bus). In such event, a charger would try to produce the same peak output current into the short circuit. After the delay, the charger would output its current limit value. If a charger is filtered, however, the filter capacitors would discharge quickly into the short circuit, and there would be a risk of clearing the dc output fuse. The risk would be especially high for a battery eliminator filter. If this would happen, the charger would be out of service until the fuse would be replaced.

AT PRODUCTS   4.4.2

The current limit is specified to be adjustable from 50% to 110% of the rated output. These limits are fixed by the control program, so you can't exceed them the way you might in the SCR/SCRF product line. Also, the response time is faster, so that peak output current is reduced.

The limited output current for HindlePower's AT charger means that a sudden load demand, such as a motor starting, requires the battery to pick up the entire incremental load. The AT charger has another feature, called a current crowbar, that protects the charger in case of a fault on the dc bus. If a fault causes an exceptionally high output current, the charger shuts down instantaneously, and restarts using its built-in soft start, or walk-in. If the current crowbar should be activated, for example, by starting a dc motor, there would be at least a four second delay before the charger would be able to contribute current to the load again.

With a dc bus fault, the AT provides its current limit value into the short circuit. The AT is designed to survive a bus fault without clearing any output breakers or fuses, so that human intervention isn't needed once the fault is cleared.

Why are there differences in the transient responses between the SCR/SCRF and AT? 4.4.2.1

The SCR/SCRF control circuit is an analog design. The response time is a trade-off between accuracy and stability. The delay was not a design objective, but it does give a little extra oomph for motor starting.

The control circuit in the AT charger is a digital design, so the current limit is controlled by software and tuned for fast response. It includes the current crowbar feature mentioned earlier that provides extra protection against damage due to a dc bus fault.

UTILITY CHARGERS 4.4.3

The UMC (Universal Maintenance Charger) and RMC (Railroad Maintenance Charger) are based on SCR/SCRF technology and have the characteristics described above. The RMC, however, adds a short-circuit detector feature because of the clamp-type battery connectors. If the output leads are accidentally shorted, the charger shuts down at once; the current limit circuit is out of the picture as long as the output is shorted.

SC CHARGER 4.4.4

The SC (Single Cell) charger is based on an SMPS and is limited to 100% of its rated output (10 Adc). You can select lower output current ratings of 3 Adc or 6 Adc, but these settings are resistively (not electronically) controlled, and the current limit may be "softer."

IF CURRENT LIMIT DECREASES WILL THIS REDUCE INPUT CURRENT?

YOU MIGHT THINK that if you decrease the current limit setting on a charger, say to 80%, that the input current would be also reduced. You would be mostly right, but the output and input currents don't correspond exactly. This is because the input current changes if the ac input voltage changes, and it also changes when you put the charger in the equalize mode. But the relationship is there, and it's used in the UMC charger to allow you to use the charger on a branch circuit that has limited capacity. By reducing the output current limit, you can reduce the ac input current below the rating of the branch circuit breaker feeding the charger. Neat? Buy one today.

WHAT HAPPENS TO CURRENT LIMITING IF CHARGERS ARE IN PARALLEL?

AS YOU KNOW, you can install an option to force two chargers to share load current equally. If you have that option in the SCR/SCRF charger, the chargers try to share the load current all the way to current limit, in what is known as peer-to-peer sharing. But because the transfer characteristics (tech talk for how well the chargers are matched) aren't exactly the same, the chargers may not share the current equally in "hard" current limit; that is, when the battery voltage is seriously depressed.

The AT charger uses a different method of current limiting, so the two chargers maintain equal current even into hard current limiting.

Without load sharing, it's probable that one charger will hog the major part of the load until it reaches current limit. At that point, the second charger will join the effort, and start to provide significant current.

TEMPERATURE EFFECTS

THIS SECTION DISCUSSES how temperature affects batteries and chargers, and discusses charger features that can stabilize charging when temperature fluctuation occurs. For simplicity, this section discusses only the features from one manufacturer, HindlePower. However, you can apply the principles discussed when considering similar options for any manufacturer's products.

HOW DOES TEMPERATURE AFFECT STATIONARY BATTERIES?

STATIONARY BATTERIES ARE expected to run at normal room temperatures for their whole lives. For this reason, batteries are designed to operate best within a limited temperature range around room temperature, normally specified as 20 °C or 25 °C (68 °F or 77 °F). Their capacity and life specifications are based on that environment. But what happens if the temperature varies? We'll discuss some examples below.

BATTERY CAPACITY INCREASES
AS TEMPERATURE INCREASES   5.1.1

Stationary battery capacity in Ah is normally specified (at room temperature) for the eight-hour discharge rate. The battery is discharged at a constant current, chosen so that the battery reaches its end-of-discharge voltage in exactly eight hours. For lead-acid, that's 1.75 VPC (volts per cell), and for NiCd, somewhere around 1.0 VPC. For example, a 100 Ah battery can be discharged at 12.5 A for eight hours. At the end of that time, the terminal voltage of a lead-acid cell will be 1.75 VPC. For a 60-cell battery, that's 105 Vdc.

NiCd batteries are useful to about 1.1 VPC. At this point, if the discharge continues, the terminal voltage collapses quickly. A 92-cell battery, at the end of its useful discharge, will measure about 101 Vdc.

Battery capacity increases at temperatures above room temperature. But before you start installing heaters in the battery room, please note the downside to life and maintenance.

LIFE DECREASES WITH ELEVATED TEMPERATURE   5.1.2

For every 10 °C in elevated temperature, a lead-acid battery's life decreases by 50%. If your battery is rated for a 20-year life at 25 °C, then at 35 °C (95 °F) it will last only 10 years. And that's assuming you use a temperature-compensated charger (see *Temperature Compensation* in Section 5.3). Without temperature compensation, there will be excessive gassing due to electrolysis (see Section 1.5.2.3).

ADDITIONAL MAINTENANCE IS NEEDED   5.1.3

If you have a flooded type lead-acid battery, more maintenance is required, especially water replenishment. There is a double danger here: If the electrolyte level falls too low, plates may be exposed, making the active material that's exposed to the air unusable. Also, the specific gravity of the electrolyte increases, accelerating grid corrosion. The only bright spot, if there is one, is that the increasing specific gravity will reduce the float current (by raising the open-circuit voltage of the cells) and slow down the effects of electrolysis.

Sealed lead-acid batteries will tolerate a moderate amount of overcharge over a limited range of temperature. But if your environment falls outside this range (above about 86 °F), the battery may experience a permanent loss of electrolyte, with additional corrosion, loss of capacity, and compromised life expectancy.

At the other end, as you might expect, you get a boost in battery life at lower temperatures. But, discharge capacity is reduced. If you've sized your battery to handle the expected loads only at room temperature, you may run short during a power emergency. Also, if you are using a non-compensated charger, there is a risk of undercharging, which over the long term may lead to permanent damage to the battery.

We noted above that flooded batteries, if operated outside the normal temperature limits, may require more frequent watering, due to gassing. Electrolyte loss can also cause additional corrosion at cell interconnections, and deposits of corrosive materials around vent caps. Evidence of corrosion can also occur for VRLA and other sealed batteries. Corrosion at terminals and connectors must be cleaned to maintain peak battery performance. There is also a risk of ground faults from large accumulations of electrolyte spray. Batteries with flame arrestor vents may clog due to excessive gassing.

NOTE: Battery manufacturers recommend cleaning solutions for corrosion (sodium bicarbonate for lead-acid, boric acid for NiCd). Be sure to use the correct solution for your battery type. Never let the cleaning solution get into the active electrolyte. Be sure to follow the manufacturer's maintenance manual for your battery.

DISCHARGE CAPACITY DROPS AT LOW TEMPERATURES   5.1.4

One of the noted benefits of NiCd battery technology is improved low temperature operation compared to lead-acid batteries. At 0 °C, a NiCd battery might have 90% of its room temperature capacity, while lead-acid would be down to 80% (lower at high discharge rates). You might be tempted to say (about lead-acid), "Well, that isn't so bad." But, as usual, there's another fly in the ointment. One of the reasons that lead-acid performance deteriorates at low temperatures is that the electrolyte's specific gravity decreases as the battery discharges. This reduces the activity of the electrolyte, leading to higher internal resistance and reduced capacity. *It also raises the freezing temperature of the electrolyte.* A fully charged battery won't freeze solid at any temperature you're likely to encounter, unless you work in Antarctica in the winter. But a discharged battery will have a freezing point above 0 °F (-17 °C) – still mighty cold, but possible in many areas for a battery in an unheated building.

NOTE: It's true that specific gravity increases as the temperature decreases, but so does the viscosity. That makes the electrolyte less mobile, contributing to the decrease in capacity.

The specific gravity of the electrolyte in NiCd batteries doesn't change as the battery discharges, so the freezing point doesn't change with the state of charge.

HOW DOES TEMPERATURE AFFECT CHARGING?   5.1.5

Just as temperature changes affect battery discharge characteristics, they also affect charging characteristics. High temperatures, in particular, can have a negative impact on battery performance and life.

Lead-acid and NiCd batteries both exhibit a negative on-charge temperature coefficient. That means that as the battery temperature rises, the battery terminal voltage decreases if the charging current is kept constant. It doesn't matter if the temperature rise is due to increasing ambient temperature or internal heating caused by the charging process. In the other direction, the terminal voltage rises if the temperature goes down.

KEY CONCEPT

Since virtually all stationary chargers are the constant-voltage type, this means that as the temperature increases, the float current also increases. Increased float current can have serious negative effects on battery life, as we described above. If you want to keep the charging current constant, you must decrease the charging voltage as the temperature rises.

Since we normally can't measure the float current, we depend on the battery manufacturer's recommendation for float voltage, which is usually specified at 25 °C or 20 °C. The float voltage is a function of battery type and construction; for lead-acid batteries, it also depends on specific gravity. The float voltage recommendation is a balance between rated capacity, life, and maintenance requirements. Once we know the best float voltage at 25 °C, then we need to adjust the voltage for the actual battery temperature, either manually (you don't want to do that), or automatically with a temperature-compensated charger.

A temperature-compensated charger is designed to adjust the output float voltage (equalize, too, of course) automatically as the temperature changes. You can read more about specifying and using temperature compensation in SECTION 5.3.

Also, there are additional descriptions of temperature effects on discharging and charging batteries in SECTIONS 1.4.6 and 1.5.6.

HOW DOES TEMPERATURE AFFECT BATTERY CHARGERS?

COMPONENT RELIABILITY   5.2.1

Filter Capacitors   5.2.1.1

A filtered battery charger uses aluminum electrolytic capacitors in the dc output filter. Capacitors typically have a maximum service temperature of 85 °C. This doesn't mean that it's OK to put the charger in a hothouse. Like other practical components, capacitors have self-heating effects: They get warmer than the environment just by doing their job. Since the charger is rated to operate in a 50 °C environment, and has an internal cabinet temperature rise of 5 °C to 10 °C, this gives us a comfortable margin of about 25 °C. The expected capacitor temperature rise is between 10 °C and 15 °C, including the internal cabinet rise.

Low temperature operation doesn't limit the life of electrolytic capacitors, but the capacitance may be reduced, leading to higher ripple voltages in the dc output. At higher temperatures, there's no significant change in capacitance or internal resistance, but life is reduced. Here's that pesky number again: For each 10 °C increase in temperature, life is reduced by 50%. Cooler is better.

Carbon Resistors (SCRF Series)   5.2.1.2

The dc output voltage of the charger is controlled by a feedback circuit that uses carbon resistors to sample the output voltage. Their resistance usually decreases as temperature rises. For a non-temperature-compensated charger this could lead to undesirable changes in the output voltage. The sampling circuit is *potentiometric*, dependent on the ratio of two resistances instead of the value of one resistor. This means that any change in the sampled output voltage depends on how well the values of two similar resistors track as the temperature changes. As long as the resistors are in the same temperature environment, the deviation in output voltage will be within the charger's ±0.5% specification.

Circuit Breakers   5.2.1.3

Standard thermal-magnetic circuit breakers are calibrated for rated performance at 40 °C. This means that they will carry their rated trip current forever at 40 °C ambient temperature. The 40 °C value is mandated by UL in the circuit breaker standard, UL489.

Above 40 °C, the trip current of a thermal-magnetic breaker may be reduced. The trip value at 50 °C may be reduced by about 10%. Because of this, one manufacturer specifies circuit breaker ratings according to the NEC requirement that the breaker must be rated no lower than 125% of the expected load current; that is, the breaker will carry no more than 80% of its rated trip current. This provides a margin of about 10% to prevent nuisance trips.

If you expect your application to spend a lot of its lifetime above 40 °C, circuit breakers can be supplied custom with high temperature calibration.

HOW CAN I SAFELY USE A BATTERY CHARGER OVER ITS 50 °C RATING?   5.2.2

A battery charger can be operated at its full power rating up to 50 °C. Obviously, it would work at 51 °C, and for a range of temperatures above that. But what happens to it? Why is there an upper limit? As we noted above in the section on filter capacitors, electronic components have upper temperature limits; beyond those limits, performance degrades, and there may be permanent deterioration, component damage or catastrophic failure, which you probably wouldn't like.

Fortunately, there is a way to use a charger at elevated temperatures, by derating the output current. Operating at a lower output current reduces the stresses on temperature-sensitive components, which allows you to use the charger in environments above 50 °C. The absolute temperature limit, though, is 70 °C, since that's the upper limit for some internal components. See the section on derating in Appendix B.

> Note: Agency certifications, such as CSA (Canadian Standards Association), are based on a maximum operating temperature of 50 °C. If you run a charger beyond this temperature, even with derating, you will be exceeding the parameters of the certification. Just so you know.

WHEN IS IT USEFUL TO INCLUDE AN OVER-TEMPERATURE ALARM? 5.2.3

Well, by now you know that letting things get too hot is a no-no. You're probably asking yourself, "How can I keep tabs on this charger, to make sure that nothing goes wrong and causes it to overheat?" I hope you are.

There are lots of ways that a charger could overheat. If you have a fan-cooled model, the fan could become blocked by dust or debris, or simply wear out. Unfortunately, they do; fans are probably the least reliable part, which is why an over-temperature sensor is supplied with a fan-cooled, or *forced convection*, model. Other causes of overheating could be blocked heat sinks, blocked cooling vents, or installing other equipment too close, resulting in inadequate air space around the charger. Failure of an internal component or maladjustment of a setting such as current limit could also cause overheating.

If you're concerned about overheating in a natural convection-cooled charger, you can order an optional over-temperature alarm. The alarm uses a thermostat mounted on one of the semiconductor heat sinks. In case of an over-temperature condition, the thermostat, which is a normally closed bimetallic switch, opens, triggering an alarm. The alarm annunciator may be a local audible or visual indicator, or a relay for remote indication.

What do you consider to be too hot? 5.2.3.1

In a charger that's cooled by forced convection, any heat sink over 70 °C is too hot. In natural convection, we trigger the alarm at 100 °C.

Why is there a difference? Sounds like it should be 70 °C all the time. 5.2.3.2

In a fan-cooled charger, a heat sink over 70 °C probably means that a fan has failed, and the charger needs attention. While we could allow the heat sink in a fan-cooled unit to run at a higher temperature, it's designed to have a lower temperature rise, to take advantage of the fan. Running it long term without the fan could spell disaster. If the alarm signals, you need to service the charger.

> Couldn't we just use the over-temperature relay to
> turn off the charger when it gets too hot? Then it
> would go back on when the charger cools down.   5.2.3.3

I'll pretend you didn't ask that. First, the charger still has a problem; shutting it down doesn't fix it. Second, and more important, you will be cycling the battery repeatedly, which could drastically shorten its useful service life. Third, the thermostat may be only a signal-level contact and can't handle the power required by the charger.

IS THERE A LOW TEMPERATURE LIMIT FOR
CHARGER OPERATION?   5.2.4

A charger operates fine down to temperatures well below 0 °C. The charger specifications are rated over a range of 0 °C to 50 °C; thus, characteristics like voltage regulation may not be within their guaranteed values outside this range.

Most internal components are rated for operation at -20 °C, and some at as low as -40 °C. Although I've never seen it happen, it's possible that SCRs might not turn on at extremely low temperatures.

Optional cabinet heaters are available. While these are primarily designed to prevent condensation damage during storage, they may help if performance is degraded in the big chill. We also urge you to acclimate a charger to a warm room for 24 hours before energizing, if it has just come in from the cold.

HOW SHOULD I STORE A CHARGER AT LOW
TEMPERATURES?   5.2.5

Those heaters that we mentioned in the previous paragraph won't necessarily keep you toasty, but if the charger is to be stored in an unheated building, they'll prevent condensation from forming potentially damaging ice crystals inside delicate components. Heaters are available for operation from 120 Vac or 240 Vac and have their own circuit breaker.

The heaters are thermostatically controlled to turn off at about 21 °C (70 °F). We do not recommend outside storage, but if you have to do it, keep the charger in its original packaging, and energize the heaters.

TEMPERATURE COMPENSATION

WHAT TEMPERATURE COMPENSATION OPTIONS ARE AVAILABLE?   5.3.1

HindlePower's AT Product Line (AT10.1 & AT30 Battery Chargers)   5.3.1.1

All AT type battery chargers use an external temperature probe, mounted on or near the battery, for temperature compensation. The same probe works with either lead-acid or nickel-cadmium batteries. When you order the option, you specify the length of the cable linking the temperature probe with the charger. Cables are available in 25, 50, 100, and 200 ft lengths (7.5, 15, 30, and 60 m).

HindlePower's Original AT10 Battery Charger   5.3.1.2

Temperature compensation probes are not compatible between the older AT10 charger and the newer AT10.1 models. When you order temperature compensation for the older AT10, you have to specify the model number, dc output voltage and battery type. A 25-ft cable is supplied as standard, but longer cables are available.

HindlePower's SCR/SCRF Battery Chargers   5.3.1.3

SCR type chargers are normally supplied with an internal temperature probe (mounted in the charger enclosure) for sensing the ambient temperature in the space where the charger is installed. Please specify the dc output voltage and the battery type to be monitored.

HindlePower's SCR/SCRF Battery Chargers with External Probe   5.3.1.4

An external temperature probe is available but requires additional components to be installed in the charger. Various cable lengths are available. Order according to battery type, charger output voltage, and cable length required. You must also order the High DC Voltage Shutdown option.

HOW CAN I MOUNT A TEMPERATURE PROBE RELIABLY?   5.3.2

The temperature probe (see Figure 5a) is designed to mount "on or near the battery." But charger instruction manuals caution against mounting on plastic battery jars. This may cause some confusion. Reprinted in the boxout is an *Application Note* that we hope will clear up the confusion:

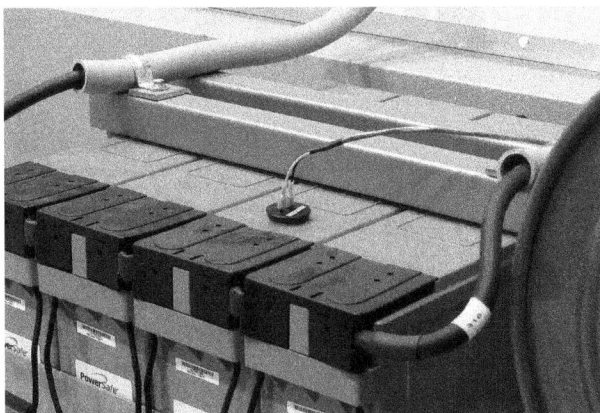

Figure 5a: A temperature probe mounted on a battery for use in temperature compensation

APPLICATION NOTE

When the AT series temperature probe was first designed, we were advised that the mounting adhesive might not be compatible with some plastic materials, including some used in manufacturing battery jars. To avoid problems, we advised staying away from all plastics as a mounting surface. We recently researched the adhesive again and can provide some additional guidance.

The mounting tape used for the temperature probe is a 1 mm (0.045") thick neoprene foam, coated on both sides with an acrylic adhesive. The adhesive is compatible with a wide range of materials, including many plastics. It also maintains strength at elevated temperatures (up to 100 °C, although we hope your battery never gets that hot!). The neoprene foam conforms to irregular surfaces, allowing the probe to be mounted on rough or coarsely painted surfaces. The probe is electrically insulated, with at least 500 V isolation.

The adhesive is optimized for high surface energy materials (most metals, for example). Thus, bare wood and galvanized steel are still no-nos, although untreated zinc is all right. The adhesive is compatible with many polymers used in battery manufacturing. However, some polymers have very low surface energy, and reliable adhesion may be difficult to achieve. Those are the materials in the "Poor Adhesion" column in the table below. The materials in the "Good" column should provide good adhesion. Most painted surfaces are OK, but you need to know that the paint will stick to the surface like, well, paint.

We still recommend an intercell connector as the best place for mounting the probe. In a battery room with little air movement, the connector will be closer to the actual battery internal temperature than the outside of a battery jar.

Table 5a contains a list of compatible materials:

Table 5a: Adhesion of temperature probes for various materials

Good Adhesion	Poor Adhesion, but Compatible
Most metals	PVC
Glass	Acrylic
Nylon	Polystyrene
Epoxy paint	Polyethylene
Polyurethane paint	Polypropylene
ABS	-
Polycarbonate	-

UNDER THE HOOD: HOW DOES TEMPERATURE COMPENSATION WORK?   5.3.3

What is temperature compensation? If you've read Section 5.1, *How Does Temperature Affect Stationary Batteries?*, you know that temperature changes can affect charging requirements. Temperature compensation is a feature of a battery charger that automatically adjusts the dc output voltage of a charger to provide just the voltage the battery needs at any temperature – that is, the voltage that will maintain the charge (float voltage). The goal is to keep the float current constant.

KEY CONCEPT

Why do we need to keep the float current constant? Stationary batteries are typically float charged for very long periods of time. We normally choose the float charge current so that it's just enough to overcome the self-discharge rate of the battery. Any less, and the battery slowly self-discharges, and doesn't have the capacity needed when it's called upon to do its job (such as operating circuit breakers in a power emergency). Any more, and the excess current (excess ampere hours over time) causes electrolyte loss, grid corrosion, and other nasty things. The electrolyte loss is due primarily to *electrolysis*, which breaks up the electrolyte into hydrogen and oxygen. The hydrogen is explosive (Hindenburg syndrome), especially if it ends up in a confined space. The oxygen is generated at the positive plates, where it can cause corrosion, reducing the life of the battery. In sealed lead-acid (VRLA) batteries, the hydrogen and oxygen are normally recombined internally back into water, replacing the electrolyte that would normally be lost during extended charging.

Many years ago, battery manufacturers empirically determined the on-charge temperature coefficients of secondary batteries. The coefficients are functions of battery chemistry, and not physical characteristics such as flooded vs. "sealed" construction, or differences in grid alloys. Therefore, you won't find any differences between, for example, antimony or calcium or selenium alloys for lead-acid, or between sealed and flooded NiCd batteries. With modern graphical tools, manufacturers can provide precise mathematical expressions for determining the correct float voltage at any temperature.

There isn't a lot of consistency in the published values for temperature coefficients, for a variety of reasons. They may be determined at different float voltages, since the change in mV per cell is naturally a function of the float voltage. Some of the original measurements for lead-acid were done at a charge voltage of 2.4 VPC, since that was considered the threshold of gas evolution, and resulted in a temperature coefficient of -3.33 mV per cell per degree F. For example, in a 60-cell lead acid battery, the on-charge voltage would decrease by about 0.2V for each °F increase in temperature (while keeping the float current constant).

The output voltage of a battery charger is normally "flat." That is, it doesn't vary with temperature, pressure, relative humidity, or the phase of the moon. This would be fine in a perfect world, where your battery would be in a temperature-controlled room, with an air conditioner that was never down for maintenance. But let's face it: That's a pretty tall order.

In Section 5.1.5 on the temperature effects on batteries during charging, we learned that if you keep the float voltage constant, float current increases as the temperature goes up. That increase in float current, over the long haul, results in excess ampere hours of charge, causing electrolyte loss and grid corrosion, shortening battery life. If you have a VRLA battery, it will be tolerant of higher float current over a limited temperature range (about 15 °C to 30 °C); if you have a flooded battery, you can always add water to replenish lost electrolyte. But you're still stuck with the grid corrosion, and if there's a lot of gassing, there will be an excessive amount of active material knocked off the grids. Remember, that's why there's extra space at the bottom of the battery. Why take a chance?

There's another factor with VRLA batteries. If you can guarantee that the battery will *never* be in a room above 30 °C (86 °F), you will be OK. But if your VRLA battery is mated to a non-compensated charger in a hot environment, you risk thermal runaway, wherein the increased float current causes internal heating, which raises the temperature, increasing the float current, etc., until you literally have a meltdown.

If you don't use temperature compensation, and you keep your charger set to the float voltage specified for 20 °C, the float current doubles for each 10 °C increase in temperature. This results in increased electrolysis (loss of electrolyte) and requires more frequent addition of water. This is OK if you have a flooded cell type, but what if you have a VRLA, and you can't add water? A VRLA will actually recombine the hydrogen and oxygen within a limited temperature range, as long as the gas evolution rate is within the design capability of the cell. If the gassing rate is too high, the cell vents the gas, and that means water loss: the cell begins to dry out, losing capacity.

On the cold side, the charging voltage should be increased if the battery is exposed to temperatures below about 15 °C (59 °F) for an extended period, to avoid undercharging. Undercharging sacrifices capacity, at the least, and undercharging for extended periods causes irreversible damage to the negative plates of a lead-acid battery.

Manufacturers sometimes specify the on-charge temperature coefficient for a battery, but there isn't a lot of consistency. Based on history and research, HindlePower settled on a temperature compensation coefficient of -3 mV/Cell/°F for lead-acid (at about 2.2 VPC), and -2.7 mV/Cell/°F for NiCd (at about 1.42 VPC).

The *slope* (in percent) of the on-charge voltage should be the same at other float voltages. We translated the coefficients to values useful for electronic circuit design: -2.5 mV/V/°C for lead-acid, and -1.9 mV/V/°C for NiCd. By using mV/V (essentially the same as a percentage), the slope becomes independent of actual float or equalize voltages. So the factor of -2.5 gives us -3.33 mV/cell/°F at a standard voltage of 2.40 VPC, but at 2.25 VPC the slope is 3.125 mV/cell/°F.

The original design for the temperature compensation probe in the SCR/SCRF products used a thermistor network, in order to make the probe characteristics consistent in manufacturing. Installing the probe in the charger alters the feedback control circuit to achieve the proper voltage/temperature response. This is also the method used in early models of the AT product line. There is a downside to this method: Battery voltage is applied to the thermistor probe, and if the probe is mounted remotely (such as on the battery), damage to the probe wiring could cause catastrophic failure of the wiring and potential loss of charger voltage control.

The current AT design uses an arithmetic algorithm to calculate the required float voltage, but still uses a thermistor probe. However, accidental damage to the probe wiring has been fail-safed. In the event that a temperature probe would be damaged or the connection to the probe would be lost, the charger would revert to linear voltage control and display an error warning the user of probe failure.

In the AT series charger, the front panel voltmeter always displays the recommended float voltage, regardless of battery temperature. The actual output voltage of the charger will be different, and is a function of the current temperature at the probe.

In the SCR/SCRF charger, the front panel voltmeter displays the actual charger output voltage, which is adjusted according to the temperature at the probe. The reference temperature for temperature compensation is 25 °C. If the manufacturer specifies your battery's float voltage at 20 °C, you need to adjust the charger output voltage for the difference. The best method is to disconnect the probe temporarily (reconnect the feedback resistor R2 into the circuit) and set the float voltage. Then reconnect the temperature probe. Alternatively, you can adjust the output voltage according to the graph shown in the charger instruction manual, but this isn't as accurate.

ALARMS & OTHER OPTIONS

YOU CAN INSTALL a "bare bones" charger at your site, and it will take care of your battery just fine. However, for some applications, you may need a charger that has additional features. Chargers can do more than simply charge a battery. Options are available that enable the charger to monitor its own performance or the characteristics of the dc bus, and transmit exceptions data using isolated relay contacts, or within a SCADA system, or even across the internet to a remote monitoring site.

This section describes various alarm circuits and options. It will help you understand how alarms work, and which alarms make sense for your site and application. Examples are given from one manufacturer, HindlePower, for simplicity but you can apply the principles discussed when considering these options for any manufacturer's products.

First, we should explain how alarm relay contacts are identified in wiring and schematic diagrams.

HOW ARE RELAY CONTACTS SPECIFIED?

SOME MANUFACTURERS SHOW alarm relay contacts *in the non-alarm condition* as their document standard for schematic and wiring diagrams. Therefore, the contact designations you see on schematic diagrams may not be the same as the "on the shelf" (non-energized) relay designations.

In many applications, alarm relays are normally energized, so that a loss of power to the alarm circuit causes an alarm as a "fail-safe." This is an essential protocol, for example, in an alarm that is intended to signal a low dc bus voltage. If power to the alarm circuit were to fail, an alarm signal would bring someone running to investigate the problem, just as if the dc bus voltage were too low. Presumably, that person would then fix the problem in the alarm.

Not all alarm relays are normally energized, though. As an example, consider the CASM (Combined Alarm & Status Monitor) alarm PC Board (in the SCR/SCRF charger). Part of the schematic diagram is reproduced in Figure 6a, showing the alarm relay terminal block and the labeling for the user's external wiring.

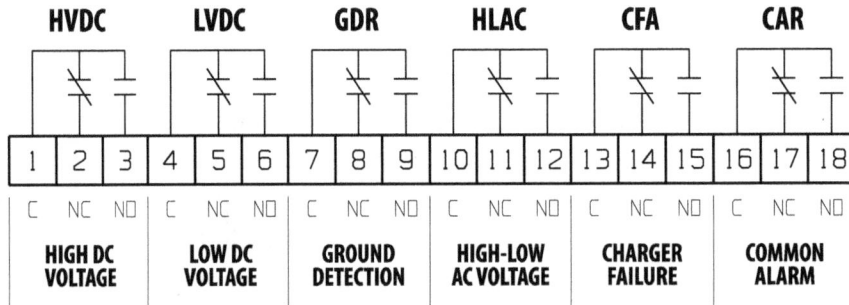

HVDC			LVDC			GDR			HLAC			CFA			CAR		
1	2	3	4	5	6	7	8	9	10	11	12	13	14	15	16	17	18
C	NC	NO	C	NC	NO	C	NC	NO	C	NC	NO	C	NC	NO	C	NC	NO
HIGH DC VOLTAGE			**LOW DC VOLTAGE**			**GROUND DETECTION**			**HIGH-LOW AC VOLTAGE**			**CHARGER FAILURE**			**COMMON ALARM**		

Figure 6a: Output connections for CASM alarm PC board in SCR/SCRF charger

The markings for the terminal block clearly show the arrangement of relay contacts – in the order C (common), NC (normally closed), and NO (normally open) – for every relay in the option. The designations NC and NO refer to the *non-alarm state* of each relay. For the Low DC Voltage alarm, for instance, the terminals C and NO mark the contact that is open as long as the dc bus voltage is normal. If the bus voltage falls below the alarm point, the NO contact closes, and the NC contact opens.

The Low DC Voltage alarm relay is normally energized, so you know that the NC and NO contacts shown here don't correspond to the labeling on the relay itself. By contrast, the High DC Voltage alarm relay isn't normally energized because you can't have a high dc voltage without ac power present, so we wouldn't want an ac power failure to cause an erroneous High DC Voltage alarm. In this case, the terminal block is labeled with NC and NO as they appear on the non-energized relay.

The labeling on the terminal block in each case eliminates any ambiguity about how to wire to your alarm. The person installing the wiring doesn't need to know anything about the relay or its state of energization, or about how the alarm circuit works.

HOW DO I UPGRADE RELAY CONTACT RATINGS FOR MY CHARGER?

THE RELAYS USED in the standard alarms have contact voltage and current ratings that are adequate for most low-level signaling applications. Table 6a shows contact ratings for the most common alarms:

Product	Contact Rating for AC	Contact Rating for DC
SCR/SCRF (CASM Alarm PC Board)	120 Vac, 0.5 Aac	120 Vdc, 0.5 Adc
SCR/SCRF (Legacy Alarms)	120 Vac, 1.0 Aac	110 Vdc, 0.1 Adc
AT Series (Standard Alarms)	120 Vac, 0.5 Aac	110 Vdc, 0.25 Adc

Table 6a: Relay contact ratings for most common alarms for several types of chargers

These ratings are usually high enough for signal-level remote annunciation, but you may need a higher rating – for example, to operate a local indicator light, or an external user-supplied contactor.

You can order auxiliary relays with higher contact ratings. Octal-based plug-in relays are available with contact ratings of 5.0 A at 120 Vac, and 0.5 A at 130 Vdc. Need higher ratings? Relays with ratings up to 10 A at 150 Vdc are available even if they're not listed in your manufacturer's catalog.

HANDLING INDUCTIVE DC LOADS   6.2.1

If you want to drive an external dc relay, using the standard on-board alarm relay (in the CASM, e.g.), you need to handle the inductive energy from the external relay coil when the CASM contact opens. The inductive "kick" caused by the external relay turning off causes arcing in the CASM relay contact, which at a minimum will shorten the relay's life. In the extreme, it could weld the contacts together and the external relay would never de-energize.

Yes, there's an app for that. The instructions for adding a freewheeling diode to handle the inductive energy are in HindlePower Application Note JD5011-00, available on the company's web site.

» *Q: Where does that "inductive kick" come from?*

  A: When you first energize the external relay, current flows in the relay coil, and that sets up a magnetic field in the relay. It's the magnetic field that does the work of pulling the relay contact closed. Magnetic fields, though, don't go quietly into the night. When you want to release the relay contact, you need to interrupt the current in the coil, but the magnetic field sees things differently — it wants the coil current to continue. So it starts raising the circuit voltage to try to keep the current flowing. When the voltage gets high enough, it causes arcing in the CASM relay. The solution is to provide another path for the coil current, and that's what JD5011-00 does.

WHAT ARE THE STANDARD ALARMS?

IN THE AT charger series, a set of basic alarms is included as standard equipment. These are essentially the same alarms that are provided in the SCR/SCRF charger with the optional CASM (Combined Alarm & Status Monitor) alarm board.

The major difference between the two products is that the CASM has an individual relay for each alarm, along with the front panel indicators, while the AT charger doesn't have separate alarm relays, but does include a Common Alarm relay. Like the CASM, the AT charger has a front panel indicator for each alarm condition.

An optional Auxiliary Relay Board is available for the AT charger, providing a separate form C relay for each alarm, plus an additional common alarm. This optional board also allows the alarms to be latched, so that the alarm signal will be persistent until someone intervenes to correct an existing fault.

HindlePower developed the CASM to offer a lower-cost path for customers who want the most commonly used alarms. That philosophy was carried over into the AT charger design. If you order a CASM or an AT charger, you probably don't need any other options to perform basic charger and dc bus monitoring.

TECH TIP

What follows are descriptions of each of the standard alarms. We also note where there are differences in behavior between the CASM and the AT alarms.

AC FAILURE 6.3.1

As the name implies, an AC Failure alarm monitors the incoming ac power, and provides an alarm when the power fails. The original alarm was simply a relay connected to the incoming power, and when the power failed, you'd get an alarm. There is no time delay for the alarm.

Fine, as far as it goes. But it's good only for a complete power failure because it depends on the sensitivity of a relay coil. In a three-phase charger, it wouldn't detect a phase failure, and it's no good for brownouts.

In the CASM design, the relay coil is replaced by an electronic circuit with adjustable sensitivity, so you can detect any level of brownout, if desired. The CASM circuit also detects a high ac voltage, and provides separate front-panel indicators for low and high ac input voltage. A single alarm relay transfers on either low or high ac voltage.

In the AT charger, only low ac voltage is detected, and the brownout level is set by the control circuit program. In both systems, hysteresis is added to prevent instability if the ac voltage is hovering around the brownout voltage. Both have time delays built in, so that a very short-term power failure won't create a nuisance alarm. And both are independent of frequency, so 50 Hz and 60 Hz work identically.

Other ac failure alarm options are available. If you like the idea of a simple relay, but want a time delay, you can order the standard ac failure alarm relay, with an added optional Time Delay Relay (of course). Now you have two relays instead of one, and you're well on your way to the cost of a CASM. This is what you need, though, if you want the time delay to be adjustable (usually from four seconds to two minutes, though other ranges are available).

If you don't have, or want, a CASM, but still need to be warned about both high and low ac voltage conditions, order the HLAC (High-Low AC Voltage alarm). Both the low and high voltage alarm points are adjustable. There is a fixed time delay of about 20 seconds.

LOW DC VOLTAGE 6.3.2

In the CASM, the low dc voltage alarm is adjustable down to 1.75 VPC (for lead-acid) or 1.05 VPC (for NiCd), and handles the number of cells normally found in industrial systems. It's normally adjusted at the factory to 2.0 VPC/1.2 VPC, but you can specify a different voltage. It has a 15-second, non-adjustable time delay.

Calibrating the low dc voltage in the field requires discharging the battery slightly, or disconnecting the battery temporarily (but only if your charger is filtered). All CASM adjustments are described in procedure JD0036, which is available on HindlePower's web site.

The alarm in the AT series has the same adjustment range, but the adjustment can be made from the front panel while the charger is operating normally, without discharging or disconnecting the battery.

Low Level Detector (LLD)   6.3.2.1

» *Q: Those alarms in the AT charger are controlled by the on-board computer. What if the computer loses power? I won't get any alarms, right?*

A: *If the on-board computer loses power, it usually means that something catastrophic has happened. In that case, the common alarm relay on the control circuit board also loses power, and it's designed for fail-safe operation. If you're monitoring the contacts in that relay, you'll get an alarm, although you might not know exactly what has gone wrong.*

The other possibility is that something naughty might happen internally on the circuit board, or even within the computer program. For that case, there's an override analog circuit, separate from the computer, that monitors the battery voltage: the LLD (Low Level Detector). If the charger stops charging, the decreasing battery voltage is sensed by the LLD, and causes the common alarm relay to send an alarm. This happens no matter what is going on in the computer.

Moral: It's a good idea to monitor the common alarm relay, no matter what other alarms you're using.

TECH TIP

End of Discharge (EOD)   6.3.2.2

There's a special case of a low dc voltage alarm: an EOD (End-of-Discharge) alarm. This is usually set at the factory to alarm when the battery voltage decreases to 1.75 VPC (for lead-acid) or 1.05 VPC (for NiCd), but you can adjust it to suit your application. This alarm is particularly important for lead-acid batteries to prevent over-discharge. You can use the alarm relay to open a dc contactor to unload the battery. This is useful at sites that lack computer control or other means to shed load automatically when the battery discharges.

TECH TIP

The EOD isn't part of either the CASM or the AT alarm package, but can be ordered separately for both systems. If you are also using a common alarm circuit, note that the EOD is never included in a common alarm. You must monitor the EOD alarm relay separately.[1]

HIGH DC VOLTAGE 6.3.3

» *Q: I know that if there's a power failure, I'll get a low dc voltage alarm. But what could cause a high dc voltage alarm?*

    A: There are a few possibilities. Unfortunately, most are related to component failure. Fortunately, they're very rare.

Of course, there's always maladjustment of the float or equalize voltage by the user. That leaves material defects, manufacturing mistakes, accidents, acts of nature, and sabotage.

- A PC Board component failure could affect circuit operation, producing high dc voltage.
- A wiring fault, or component failure in the feedback circuit, could interrupt the dc feedback signal, allowing the charger to "run away," producing a high output voltage.
- SCRs are temperature-sensitive devices. Excessive temperature could cause the SCR to lose control,[2] allowing the output voltage to rise. This might be caused by an ambient temperature that is too high, by blocked cooling vents, or fan failure.
- A very rare type of SCR failure, an anode-to-gate short, could also cause the SCR to lose control.
- Water ingress can cause all kinds of problems with electrical and electronic parts.
- So can conductive particles, such as smoke. A charger that has been exposed to fire or flood is at high risk.

Whatever the cause of a high dc output voltage, you need to know right away because high voltage is potentially very damaging to a battery. The CASM and AT charger both have a high dc voltage alarm. Additionally, the HVDC alarm in the AT charger can be

---

1 Why? Because EOD won't be activated unless there is already an ac failure, or another cause of a low dc voltage. Either of these would already have activated the common alarm relay.

2 Specifically, this is known as loss of forward blocking. The SCR turns on (conducts) at the start of each cycle, instead of when the gate signal asks for it to turn on, which is usually later in the cycle.

set to shut the charger down automatically.[3] There is a time delay built into each alarm: 15 seconds in the case of the CASM, and 30 seconds for the AT charger.

Charger Shutdown   6.3.3.1

In the SCR/SCRF charger, HVDC shutdown isn't automatic. A separate option, appropriately named HVDC Charger Shutdown, accomplishes this task. In the event of a high dc voltage, the shutdown option removes the power from the control circuit board, disabling the charger. There is a fixed time delay built into the option, so that transient voltages (such as from a load dump) won't be a problem. A separate form C relay contact is provided to send an alarm signal.

Remember that in both the SCR/SCRF and AT charger series, charger shutdown is permanent until the charger is manually restarted by a real person (except that an ac power failure and subsequent recovery can restart the SCR/SCRF charger). The circuits are designed so that in a system with two parallel-connected chargers, only the charger with high voltage is shut down.

GROUND FAULT DETECTION   6.3.4

The CASM and the AT charger both have provision for disabling the ground fault detection alarm. Why would you want to do this?

- You have a 24 Vdc or 48 Vdc system, and one terminal of the battery is grounded. In that case, you have a permanent ground fault alarm indication, which means that the common alarm relay is also permanently activated.
- You have another ground fault detector at your site, and the alarm in the charger interferes with its operation.
- You have a ground leakage problem; you know about it, and the repair is scheduled. But you want to disable the alarm because you're only monitoring the common alarm relay, and the ground fault prevents the common alarm from signaling if there is another alarm event at the site.

---

3 Normally, the charger is shipped with HVDC shutdown disabled. You can specify that it should be enabled.

What is a ground fault, and why do I want to detect it?   6.3.4.1

In high dc voltage systems (125 Vdc and higher), the battery is "floating." That is, it's isolated from earth ground. This is done for safety, so that personnel aren't exposed to a shock hazard by accidentally touching one end of the battery. A ground fault occurs when a leakage path for dc current is created from one terminal of the battery to ground. This could be caused by water ingress (in conduits, for example), poor housekeeping, damage to battery containers, or electrolyte leakage due to overcharging.

You can measure the dc resistance of a ground fault, and it can range from zero ohms up to several tens of thousands of ohms. High resistances carry low risk, but any ground fault should be addressed. A double ground fault (from both battery terminals to ground), which could be associated with electrolyte leakage, is of great concern because of a fire risk.

How do you detect a ground fault?   6.3.4.2

There are two basic methods. A small ac signal can be applied between the battery terminals and ground; a ground fault will cause a detectable change in ac current. This method has the advantage that there is no ohmic connection between the dc bus and ground. This method is normally used by stand-alone equipment installed at a site.

The second method, and the one used on most battery chargers, is to connect a dc bridge circuit from the battery terminals to ground, using very high value resistors. A ground fault causes an imbalance in the bridge, which is detected by an electronic circuit.

The bridge circuit is the method used in both the CASM and AT chargers because of its simplicity and sensitivity. Most other detection options are also based on the bridge concept.

Is there an optimum sensitivity range for detecting ground faults?   6.3.4.3

A more sensitive detector will catch a higher resistance leakage path. You might have a minor ground fault, say, 50 kΩ or 100 kΩ. That's only 2.3 mA (milliamperes) for a battery floating at 130 Vdc.

If you need to, you can ignore those high resistance faults by making the sensitivity lower. The optimum sensitivity range for a 130 Vdc bus is from about 5 kΩ to 50 kΩ (by coincidence, the adjustment range for the AT ground detection alarm).

Do all ground detection alarms have that adjustment range?   6.3.4.4

Good question. No. In general, older style ground fault detectors, using relays or indicator lights, were a lot less sensitive, and not adjustable. The CASM detector has higher sensitivity, to about 15 kΩ to 17 kΩ, but is not adjustable.

Table 6b gives the approximate sensitivity for various ground detection circuits.

Description	Part Number	Typical Sensitivity	Notes
Ground Detection Lamps	EJ0089-xx	1,000-3,000 ohms	Includes test switch
Ground Detection Relay	EJ0086-xx	300 - 1,000 ohms	
Front Panel Meter Switch	EJ0094-xx	See discussion below	x
CASM	EJ0837-xx	17 kΩ	x

Table 6b: Sensitivity for various ground detection circuits

The Front Panel Voltmeter Switch option doesn't have a sensitivity rating as such. It simply indicates a dc voltage that's a function of the ground leakage resistance. It isn't very useful below about 50 kΩ leakage resistance, but it has the advantage that it doesn't use a bridge circuit, and so the battery has no ohmic connection to ground unless you are checking for grounds.

You can estimate the ground fault resistance using the meter switch option. Download the document JD0062-00 from this address: http://www.atseries.net/PDFs/JD0062-00.pdf

The graph in this Application Note will help you determine the fault resistance. It won't be accurate, though, if you have leakage paths from both battery terminals to ground.

A ground fault indicator light has one redeeming feature: you can see from across the room that you have a ground fault. As you can see from the table, though, it would be a serious fault.

Ground detection relays, because of their high inherent hysteresis, require a reset switch to clear the alarm once you've fixed the ground fault. It's the only legacy alarm that is self-latching.

What's the risk if there are faults from both battery terminals to ground? You called that a double fault.   6.3.4.5

If the resistance paths are nearly equal, they can't be detected using a bridge circuit. And, if the resistance is low (a few hundred ohms or less), the fault can dissipate a lot of power, with resulting risk of damage or even fire. This condition is probably caused by electrolyte leakage; routine battery inspection should discover it.

What happens to the sensitivity if I have two chargers in parallel?   6.3.4.6

Another good question. If you have a CASM, or another option that uses a bridge configuration, the sensitivity is reduced. You could disable the ground fault detector in one of the chargers to restore the original sensitivity.

I'm paranoid. Can I have two ground fault detectors in my charger?   6.3.4.7

Some users do that. A popular combination is the Front Panel Voltmeter Switch, coupled with a CASM. Remember this, though: At the factory, that combination is wired so that when the voltmeter switch is used to check for grounds, the CASM ground detector is temporarily disabled. If the meter switch option is field-installed, that may not be the case.

One last question. You said that bridge detectors use high-value resistors. How high?   6.3.4.8

In the CASM, they're 100 kΩ. In the AT charger, they're 22 kΩ. Why the difference? Two reasons. The CASM is designed to work with 250 Vdc buses; the AT only goes to a 125 V bus. Also, AT series' resistor values are smaller to get a wider sensitivity range.

CHARGER FAILURE   6.3.5

Ready for a little controversy? For years, charger failure alarms were simply zero-output-current alarms. And they weren't really zero current but could give an alarm whenever the output current decreased below about 2% of the charger rating. This could be a frequent occurrence, or even a permanent condition, depending on the installation. Having zero output current doesn't *necessarily* mean the charger has failed.

The NEMA PE 5 standard differentiates between charger failure and zero current alarms and gives several helpful examples of failure causes. But the requirement for an alarm boils down to this question: Regardless of the present output current or vóltage, can the charger charge the battery when it's called upon to do so?

The CASM has an advanced "True Charger Failure" circuit that tests the charger at intervals to be sure that it's working and is connected to the battery. At each test interval, the circuit checks the charger output current. If it's less than 2%, the circuit increases the output voltage slightly, and then looks for an increase in output current. If an increase is detected, the alarm circuit is satisfied that the charger *hasn't* failed and goes back to sleep for eight more minutes. Of course, the output voltage is also returned to the proper setting.

To understand why we do this requires a little background on the problems with a zero-current alarm:

- If a charger has an automatic equalize option, such as an auto-equalize timer, the output current goes to zero when the charger switches from equalize back to float mode. This is because the output voltage setting suddenly decreases, but the battery voltage has been raised during equalize, and takes several seconds, or even minutes, to return to float voltage. A zero-current alarm gives an alarm, but the charger hasn't really failed.
- In a site with parallel chargers, using random (not forced) load sharing, one charger may be set to a slightly higher float voltage than the other. It only takes a few tens of millivolts for that charger to "hog" all the output current. The idle charger would send a zero-current alarm, even though it hasn't failed.
- Consider a site with a large battery (e.g., 400 Ah) and large charger (e.g., 100 A), but a very small standing load. The float current for that battery is less than 0.5 Adc. You would get a zero-current alarm.

The True Charger Failure mode in the CASM avoids these problems, and alarms only if the charger is really incapable of providing dc output current. If you still want a zero-current alarm, you can set up the CASM to act that way. But you must choose. You can't have it both ways.

The AT charger takes a slightly different approach. The AT control circuit monitors the bus voltage and the charger output current, and if the bus voltage decreases below the set point (e.g., float voltage), it knows that the charger should be delivering its current limit value (or darn close) to the battery. If it isn't, the control circuit calls that a DC Output Failure, and sends an alarm.

COMMON ALARM   6.3.6

The Common Alarm relay, also called a Summary Alarm, is a single relay that includes, or summarizes, all the standard alarms: High and Low AC Voltage, High and Low DC Voltage, Ground Fault, and Charger Failure. Using this alarm relay, you can monitor all the alarm conditions with a single annunciator, or a single port in a SCADA system. What you lose, obviously, is the ability to distinguish among the individual alarms; someone must visit the charger (it's lonely, anyway) to determine what event caused the alarm.

In both the CASM and AT chargers, the common alarm has a 30 second delay. You can't choose which alarms to include; they're all in there, except that in the CASM you can remove the Charger Failure Alarm from the Common Alarm.

> **But I want to monitor the Charger Failure separately and group just the Ground Fault alarms. I don't want the other alarms in there.**   6.3.6.1

No problem. Just use the Theriault connection, described in Appendix C.

TO LATCH OR NOT TO LATCH – DO YOU HAVE A QUESTION?   6.3.7

> **Just one more. Maybe. In the section on ground fault detection, you mention latching. Do all these alarm relays latch?**   6.3.7.1

No, but...
CASM relays don't latch, but there's an app for that, too. The EJ1193-xx option gives

the CASM alarms the ability to latch, but it uses an auxiliary relay for each alarm. It's a bit of a wirer's nightmare.

Better to choose an AT charger with the optional auxiliary relay board. Starting in 2010, the relays on the auxiliary board could be set to latch, with no external wiring required. Just remember that a latching relay requires manual clearing (unless you have the communications option described in Section 6.5).

CONTROL OPTIONS

NOW THAT YOU'RE thoroughly alarmed, let's look at some control options. We'll look at automatic controls for equalizing batteries, forcing two chargers to share the dc output current, and, most importantly, adjusting the output voltage for temperature variations.

CONTROLLING FLOAT/EQUALIZE MODE   6.4.1

Since automatic equalize controls are standard in the AT charger, this section deals only with the SCR/SCRF.

Why do you recommend indicator lights
if you control float/equalize manually?   6.4.1.1

If you have a manual float/equalize switch (the default), it's smart to have float and equalize indicator lights. These indicators are mounted on the front panel, next to the float and equalize adjustment potentiometers. They allow you to see the operating mode from across the room (green for float, red for equalize).

Why is this smart? If you want to manually control equalize charging, you can get by with a manual switch. But what if you were to set the charger to equalize, and then leave for a weekend, or even vacation? The indicators can remind anyone that the charger is in equalize mode – which is especially important with valve-regulated lead-acid batteries.

TECH TIP

Why might a timer equalize option be even better?   6.4.1.2

A timer equalize option will return the charger to float mode at the end of a predetermined equalize period. There are manual timers, automatically started timers, and cycle (or percent) timers. Enjoy your vacation with a peaceful mind.

Manual Timer with Indicating Lights   6.4.1.3

The manual timer is run by a clock motor and can be set from one (or less) to 72 hours. Equalize charging begins as soon as you rotate the front panel dial to select the time. When the time expires, the charger returns to float, and stays there. This is a one-shot deal – to equalize again, you turn the dial again.

Advantages: Simple, reliable, guaranteed maximum equalize time. If the ac power fails during equalize, the timer picks up where it left off when the power returns.

Disadvantages: Someone must be at the site to touch that dial. A good choice for a staffed site, or one with a comprehensive battery maintenance program.

» *Q: OK. I have a manual timer. How long should I equalize my batteries?*

   A: Methinks this is a loaded question. You didn't say anything about your battery type, its size, or the site application. Here are a few guidelines.

- If you have a VRLA battery, an equalize period of zero hours is about right. In other words, don't equalize. You might think about setting the equalize voltage to the same value as the float voltage, so that if anyone tampers with the controls, the battery won't be damaged.
- In the AT charger, if you set the equalize timer to zero hours, the equalize mode is inhibited, so that even the manual equalize setting won't work.
- For a flooded lead-antimony or lead-selenium cell, you can equalize for about 8 to 24 hours, if you've had one or more deep discharges. If equalize charges are infrequent, aim for the upper number. If you've set the equalize voltage to the maximum value specified by the manufacturer, go for the lower number. Remember, in a series string, it's the higher capacity cells that are lower in voltage than the lower capacity cells. You want to equalize only long enough to bring up the charge in the higher capacity cells because you're simultaneously overcharging the lower capacity cells. When the

battery reaches the equalize charging voltage, the charge current starts to decrease. Terminate the equalize charge when the current to the battery falls to the C/20 rate.

- For flooded lead-calcium cells, equalize sparingly. If possible, measure individual cell voltages to determine the need for equalization. If possible, you should measure the voltages with the battery on open circuit (no charger, no load).

» *Q: I have a UPS battery that's a VRLA. Shouldn't that be equalized, since it's cycled a lot?*

A: *You probably have a thin plate, lead or lead-tin battery designed for high rate deep discharge. Although this battery type has some cycling capability, you probably aren't cycling it as much as you think; UPS is basically a standby application. With a healthy recombinant battery, the cells should be equalized after an extended time on the manufacturer's recommended float charge.*

Auto-Equalize Timers   6.4.1.4

If you don't want the hassle of manually initiating an equalize charge, or if you have a remote site with flooded batteries, you might consider an auto-equalize timer. The most common such timer is initiated by an ac power failure, on the theory that an ac failure results in a battery discharge. To guard against an equalization for every glitch in the ac power, most timers have a delay (10 seconds in the case of the SCR/SCRF charger) before "arming" the timer. The equalize charge starts when the ac power returns, and continues for the time selected (by the user) on the timer.

Advantages: Simple, reliable, guaranteed maximum equalize time. Hands-off operation. Useful for an unmanned site.

Disadvantages: If the ac power fails again during equalize (for the duration of the time delay), the timer is reset to the full equalize time. Multiple power failures can result in excessive equalize charging. Since equalization causes electrolyte loss through electrolysis, battery maintenance requirements are actually increased over a manual timer. You should not use an automatic timer with VRLA batteries.

Other Auto-Equalize Timers 6.4.1.5

You can also get an auto-equalize timer triggered by the charger's current limit signal. This is a second-order effect. Presumably, if the charger is in current limit, that means that the ac power previously failed, and the battery was discharged and should be equalized. The problem is, the battery could also be discharged slightly by a transient load that exceeds the capacity of the charger, and the charger will go to current limit for a short time to recover that discharge. The current limit-activated timer has about a 20-second fixed time delay that helps to overcome this disadvantage.

A percent timer, or cycle timer, equalizes the battery periodically, irrespective of any external conditions. Typically, the battery will be charged on equalize for a few hours each week. This might be good for NiCd or nickel-iron batteries, since they have a higher self-discharge rate than lead-acid. However, you shouldn't use it with recombinant NiCd batteries (designed for reduced maintenance) because of – you guessed it – the increased battery maintenance that will result.

SHARING LOAD CURRENT EQUALLY
BETWEEN PARALLEL CHARGERS 6.4.2

Any two battery chargers with the same output voltage can be operated in parallel. They don't even have to have the same current rating. Each charger will provide output current, as needed, up to its current limit setting. But the chances are that they will have unequal output currents, even if they're the same current rating. This is *random* load sharing, which means it isn't load sharing at all.

The unequal output currents are due to natural differences in the output voltages of the chargers over the load current range. For any given load current, you could adjust the two chargers so that they have the same output voltage, but as soon as the load current changes, the output voltages would no longer track. Technically, we say that the output transfer characteristics of the two chargers have different slopes.

What if you wanted the two chargers to share the load equally, no matter what the total load current is? For that, there is the *Forced Load Sharing* option (for the SCR/SCRF charger; it's standard in the AT charger, but requires an optional connection cable). With forced load sharing, the two chargers will have the same output current, within 2%, all the way up to their current limit settings. You can see that the total available output current is twice the rating of each charger.

And what if the chargers don't have the same current ratings? SCR/SCRF chargers will still share proportionally. Here's an example: Suppose you have one 50 Adc and one 100 Adc charger, and the total load is 100 A. The load is 2/3 of the total charger capability, so each charger contributes 2/3 of its rating. The 50 A charger provides 33.3 A, and the 100 A charger provides 66.6 A, for a total of 100 A. (Alright, 99.9. So sue me.)

One caveat: The AT forced load sharing requires the chargers to have the same current rating. The control program simply ignores your wishes if the chargers have different ratings.

Whoops, another caveat: In the SCR/SCRF charger, you can't successfully use forced load sharing and temperature compensation together. With temperature compensation active, the chargers won't share the load within 2%, and probably not even within 10%. See the next section for a description of temperature compensation.

Uh, we aren't done yet. For both SCR/SCRF and AT chargers, if you are sharing three-phase chargers, the phase rotation of the ac input voltage for both chargers should be the same for maximum stability.

BENEFITS OF TEMPERATURE COMPENSATION OPTIONS   6.4.3

If you've read Section 1.5.6 which discussed the temperature effects on battery charging, you know that the charging voltage of a battery decreases as the temperature increases. A charger that does not compensate for temperature will, at the least, have more maintenance requirements and possibly reduced battery life.

Temperature compensation options, available for both the SCR/SCRF and AT chargers, help to offset the effects of temperature changes by adjusting the charger output voltage appropriately. Temperature compensation is helpful for almost any battery installation, but is critical for VRLA batteries, especially in an environment that isn't temperature-controlled.

TECH TIP

For the SCR/SCRF product, the standard temperature compensation option uses a temperature sensor (probe) inside the charger, but an external probe is available for mounting on the battery. The AT charger option uses only an external probe. Full instructions for mounting the probe and operating the charger with temperature compensation are included with each option. If you specify the option at the time that you order the charger:

- SCR charger – For the internal probe, you don't have to do anything. If you get an external probe, mount it preferably on a battery intercell connector. Run the connection cable back to the charger and connect the wiring to TB45 as shown in the instructions. IMPORTANT: The wiring carries the full battery voltage. Run the wire carefully in its own conduit. Do not run it next to other power wiring. The polarity of the connections is unimportant. Please note, though, that in the SCR/SCRF series, temperature compensation probes aren't interchangeable.
- AT charger – Install the probe, preferably on a battery intercell connector. Run the connection cable back to the charger and connect it to TB8 as shown in the User's Manual. The cable carries only a signal level voltage, but for safety, and to avoid noise pickup, run the cable in its own conduit. The polarity of the connections is unimportant. AT series temperature compensation probes are interchangeable.

There is an application note, JD5003-00, for the AT temperature compensation, which you can download from HindlePower's web site.

For a discussion of temperature compensation in excruciating detail, see *Temperature Compensation* in Section 5.3.

COMMUNICATIONS OPTION (AT SERIES)

INTRODUCTION TO COMMUNICATIONS   6.5.1

In the "old days," chargers reported alarm status through relays. The relay contacts were used to trigger alarm buzzers, turn fans and indicator lamps on and off, or report status to the main annunciator panel. If analog data, such as voltage or current needed to be reported, transducers were installed in the charger to convert the desired parameter into a 4-20 milliampere or 0-10 volt signal. The transducer signals would be connected to transducer meters for remote monitoring or to Remote Terminal Units (RTUs) for data aggregation. Every alarm and transducer output required a dedicated pair of wires.

SERIAL COMMUNICATIONS BASICS   6.5.2

Serial Communications permit data to be transferred between devices with a minimal number of wires (typically three). The data is transferred via a sequential series of pulses and the amount of data transferred, or bandwidth, is only limited by the size of the pulses. These pulses can be created by varying the voltage on one end of a copper wire, or by turning on and off a light source on one end of an optical fiber.

The data to be transferred is converted to a binary bit stream. The rules for "serializing", or breaking the data into a stream of pulses, must be defined at each end of the communications channel such that the receiving end can restore the original data from the series of pulses it received.

SYNCHRONOUS VERSUS ASYNCHRONOUS   6.5.3

Devices on both sides of the communications channel need to be synchronized such that the receiving device will detect when and how to reconstruct the bit stream. Time synchronization can be accomplished two different ways, with *synchronous* or *asynchronous* data formats.

Synchronous Formats   6.5.3.1

Synchronous formats (see Figure 6b) require an additional CLOCK line signal connected between the devices. The signal on the DATA line will change on every rising or falling edge of the CLOCK line. The data line is sampled at the receiving device on the opposite clock edge (middle of pulse). Neither device needs to be configured for the data rate, and the data rate may change dynamically.

Asynchronous Formats   6.5.3.2

Asynchronous formats (see Figure 6c) do not require a CLOCK line; only a DATA line is connected between the devices. The receiving device monitors the DATA line and will synchronize or FRAME the bit stream based on a transition event. Most asynchronous formats use a falling edge after a period of inactivity as the synchronization method. The DATA remains low for one bit time and this is referred to as a START BIT.

DATA WORD

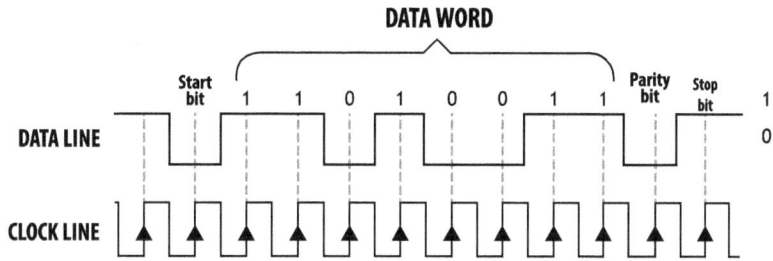

Figure 6b: Synchronous data format; requires DATA line and CLOCK line signals; when DATA signal changes it does so with rising or falling edge of CLOCK line

Devices that communicate asynchronously must be configured to use the same bit time-period (or BAUD rate). Receiving devices detect the start bit's falling edge and sample all remaining bits of the stream in the middle of bit time. Asynchronous formats are preferred because they require fewer connections. Fewer connections also reduce the amount of wire (or fiber), terminals, connectors and thereby cost.

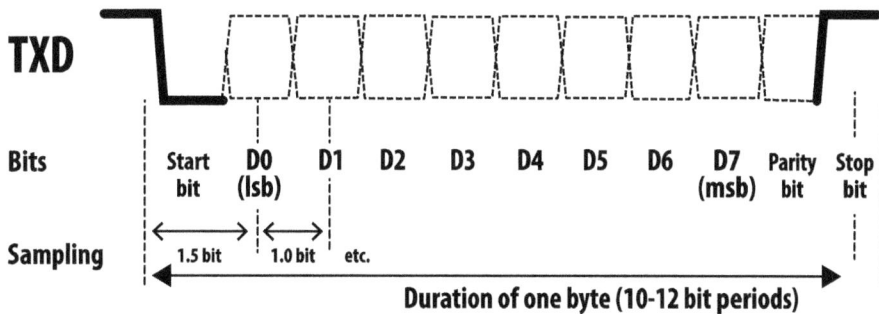

Figure 6c: Asynchronous data format; requires only a DATA line between devices

SERIAL CONFIGURATION PARAMETERS   6.5.4

Framing   6.5.4.1

Asynchronous communication information-packets are sent in frames. At a minimum, a frame consists of a START BIT, the payload (actual data bits D0-D7), and a STOP BIT. A frame may include additional error checking bits (PARITY) and/or stop bits. Devices that communicate asynchronously must be configured to follow the same set of rules. Most devices will allow these parameters to be configured to ensure operability with other serial devices.

BAUD rate   6.5.4.2

Since there is no CLOCK line between the devices, each device must send and receive data using the same bit time. This bit time is referred to as the BAUD rate. There are several standard BAUD rates for serial communication ports ranging from 110 bits per second up to 256,000 bits per second. The typical rates supported by most devices range from 9600 to 115,200 bits per second.

Number of Data Bits   6.5.4.3

The DATA BITS (D0 thru D7) are the payload of the frame. Most modern protocols transfer data in bytes (eight bits at a time). Some data protocols transfer data in ASCII which utilizes a seven-bit format. An older version of the Modbus protocol (described in Section 6.5.9) utilizes the ASCII data format.

Parity   6.5.4.4

An extra error detection bit may be added to each frame. This bit can be configured to transfer an EVEN PARITY BIT or an ODD PARITY BIT. If EVEN PARITY is selected, the PARITY BIT will be set to '1' when there is an odd number of '1's in the data bit fields (D0-D7). If ODD PARITY is selected, the PARITY BIT will be set to '1" when there is an even number of '1's in the data bit fields (D0-D7).

Stop Bit(s)   6.5.4.5

All frames must end with one (or two) STOP BIT(s). A STOP BIT is a high or logic '1' bit at the end of the frame. The STOP bit returns the bus to the idle state, which is the logic '1' state.

COMMUNICATIONS ERRORS   6.5.5

Devices with matching configuration and correct cabling will communicate without errors. The following error messages are standard for most devices with serial communications. Error messages can be helpful in determining a mismatched configuration or an application issue.

Framing Error   6.5.5.1

A framing error occurs when a device detects a START BIT, but not the STOP bit(s). It may indicate a configuration mismatch between devices for: BAUD rate, parity, number of data, or number of stop bits. This error may also indicate an unreliable connection due to excessive cable length, improper terminations, poor shielding or excessive electrical noise.

Parity Error   6.5.5.2

If parity checking is enabled, all devices receiving data will check the parity bit on every frame. Frames received with incorrect parity will generate a parity error. Most devices ignore messages received with a parity error. This error may indicate a parity configuration mismatch or an unreliable connection due to excessive cable length, improper terminations, poor shielding or excessive electrical noise.

Overrun Error   6.5.5.3

An overrun error is rare and typically only occurs if the device cannot process messages and frames fast enough such that a new frame is received before the device can process the previous frame. If this error occurs, reduce the device's BAUD rate and polling rate.

CONNECTING TO THE NETWORK   6.5.6

Most chargers support RS-232 and RS-485 networks with master/slave protocols. Ethernet connections with multiple protocol support are now becoming more common. Fiber optic versions of both serial and Ethernet networks are often implemented to provide isolation between network devices.

RS-232 6.5.6.1

The RS-232 standard was used by personal computer (PC) serial ports prior to the development of the USB standard. RS-232 provided a simple connection from the PC to mice, printers, and/or modems. The standard was designed to connect two devices together that are a short distance apart. This standard has limited cable length (typically less than 50 feet) and is susceptible to electrical noise.

The simplest RS-232 connection (see Figure 6d) consists of three wires, a TRANSMIT, a RECEIVE, and COMMON (or GROUND). The TRANSMIT of one device is connected to the RECEIVE of the other device. The COMMON of both devices are connected to each other.

RS-232 signals are inverted at both transmitter and receiver. As seen in Figure 6e, a logic "high" signal at the transmitter's input is translated to a logic "low" at the output of the transmitter. A logic "low" signal input to a receiver is translated back to a logic "high" at the output of the receiver.

The signal level for RS-232 is typically (+/-) 12 volts such that a high level or mark condition on the driver's input will result in – 12 volts on the driver's output. A low level or break condition on the driver's input will result in +12 volts on the driver's output.

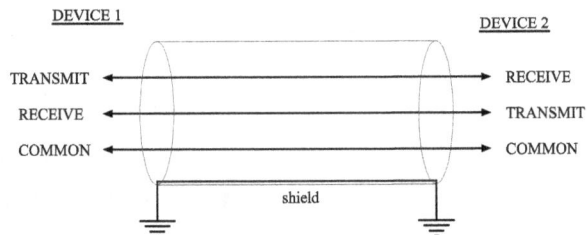

Figure 6d: RS-232 Connection between devices. The TRANSMIT of one device is connected to the RECEIVE of the other; the COMMON for both devices are connected together

Most newer RS-232 receivers will support signal levels down to (+/-) 3 volts to ensure compatibility with lower voltage RS-232 drivers. The transition between logic levels on RS-232 receivers is zero volts regardless of the threshold voltage.

Some older implementations of RS-232 supported hand shaking signals RTS (Ready-To-Send) and CTS (Clear-To-Send). This interface used five wires instead of three. The additional two wires were used to connect the RTS of one device to the CTS of the other device. The hand shaking signals were used to throttle the communications between older, slower devices to make sure the devices were ready to accept more data before the data was sent.

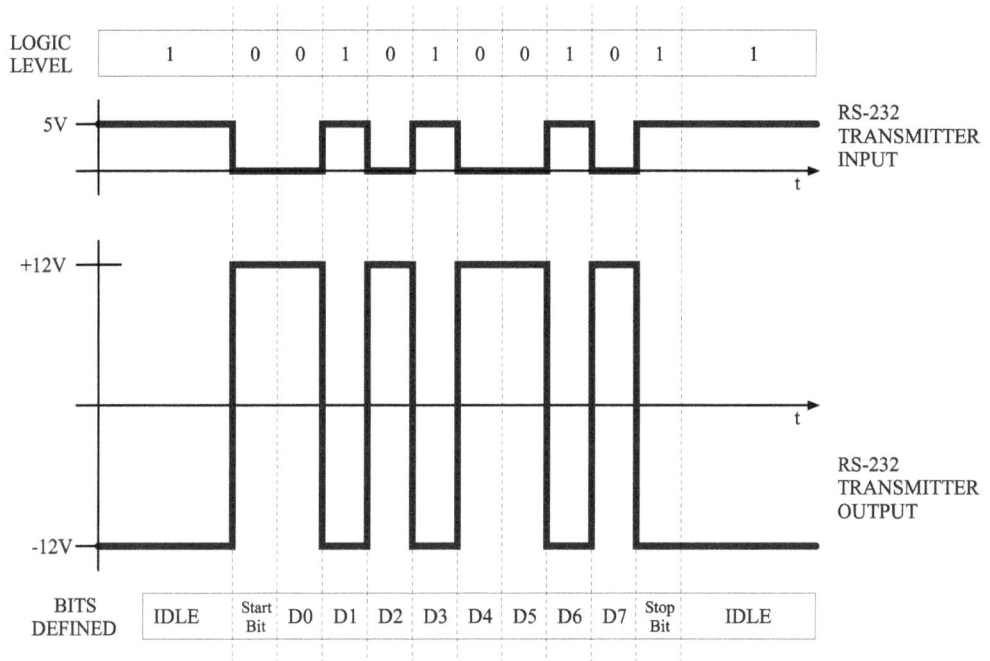

| LOGIC LEVEL | 1 | 0 | 0 | 1 | 0 | 1 | 0 | 0 | 1 | 0 | 1 | 1 |

Figure 6e: RS-232 signals are inverted by both transmitters and receivers. Diagrams show the input to a transmitter being inverted at the output of the transmitter

RS-485   6.5.6.2

The RS-485 standard defines electrical characteristics of drivers and receivers for use in balanced digital multipoint systems. RS-485 allows multiple devices to be connected to the same network (up to 32 standard devices). Balanced networks are much less sensitive to electrical noise and will support distances of 1000 meters.

RS-485 networks have TRANSMIT and RECEIVE differential connections and a COMMON (or GND). RS-485 drivers convert a single-ended TTL signal to a differential signal pair; RS-485 receivers convert a differential signal pair back to a single-ended TTL level signal. Most protocols that support RS-485 are master/slave. One master device will poll all other slave devices on the RS-485 network. Slave devices will only respond when polled.

In a 4-wire RS-485 network (see example in Figure 6f), the master TRANSMIT pair is parallel wired to all slave devices' RECEIVE pairs. All slave TRANSMIT pairs are parallel wired to the master's RECEIVE pair. The COMMON of each device is parallel wired together – though this is not pictured in the diagram below. The network typically consists of two twisted-pair cables and a common or shield connection.

The sequence of operation of a 4-wire RS-485 network is as follows:

1. The master sends a poll message directed to a specified slave device.
2. All slave devices receive the poll request.
3. The specified slave device will turn on its transmitter and respond to the poll request.
4. The slave address turns off its transmitter after the message is sent.
5. The master sends out the next poll message and the process continues.

4 - WIRE RS-485

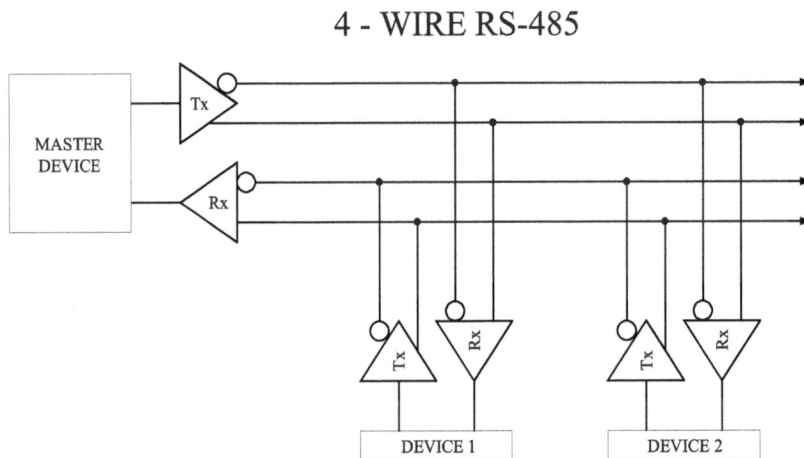

Figure 6f: Full duplex, 4-wire, RS-485 network showing two of a maximum of 31 slave devices that can be connected to the master device. The diagram shows connections for transmitting and receiving but does not show the common connections of all devices wired together.

Since there is a separate TRANSMIT pair and RECEIVE pair with 4-wire RS-485 networks, full duplex communication may be supported. Full duplex communication allows the master device to poll the next slave device, while it is receiving the response from the previous slave device's poll. If full duplex communication is supported by all devices on the network, the master device could issue the next poll during step 3 in the sequence listed above.

Full duplex operation does improve the communication bandwidth of the network since two devices can be transmitting at the same time, but not all devices support full duplex operation. Half duplex is by far the more common RS-485 configuration. In half duplex mode only one device can transmit at a time. Since only one device is transmitting at a time, only one twisted pair connection is required. All devices, both master and slave, can transmit and receive messages over the same twisted pair. This connection method is referred to as a 2-wire RS-485 network.

In a 2-wire RS-485 network (Figure 6g), the TRANSMIT and RECEIVE differential signals of each device are parallel wired to the same twisted pair. The COMMON of each device (not shown) is also parallel wired together. The network between the devices typically consists of a twisted-pair cable with a common or shield connection.

2 - WIRE RS-485

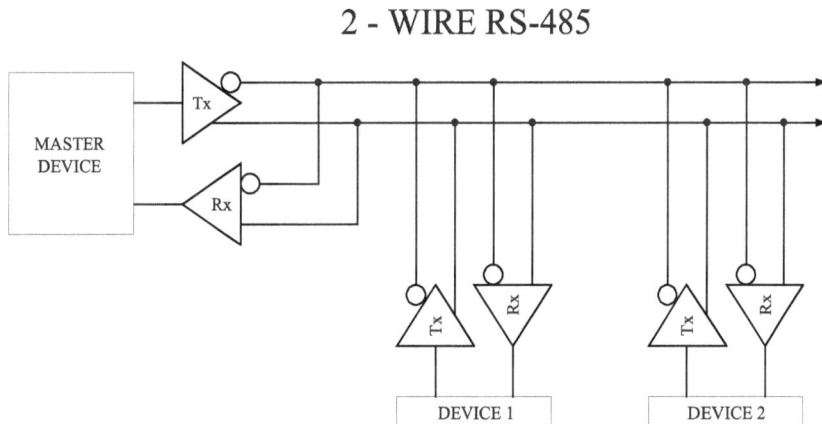

Figure 6g: Half duplex, 2-wire, RS-485 network showing two of a maximum of 31 slave devices that can be connected to the master device. The diagram shows connections for transmitting and receiving but does not show the common connections of all devices wired together.

The sequence of operation of a 2-wire RS-485 network is as follows:

1. In the idle state, all devices have their transmitter turned off.
2. The master device initiates a poll by turning on its transmitter. The master sends the poll message and turns off its transmitter.
3. All slave devices receive the poll request, but only the slave device that the message was directed to will respond.

4. The specified slave device will turn on its transmitter and respond to the poll request.
5. The slave device turns off its transmitter after the message is sent.
6. The network is back in the idle state and the master device initiates the next poll.

RS-485 Differential Signals   6.5.6.3

The differential signal pair consist of a (V+) and a (V-). The logical state of the bus is determined by the voltage difference between (V+) and V(-) as pictured in Figure 6h. These signals are referenced to each other, not to ground.

A high logic state is present when there is a positive voltage differential at (V+) when referenced to (V-).

A low logic state is present when there is a negative voltage differential at (V+) when referenced to (V-). A 200mV differential voltage is required to ensure a valid state.

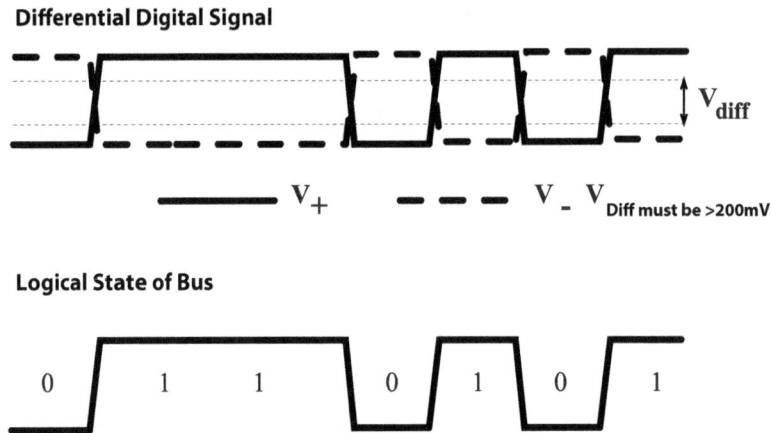

Figure 6h: In an RS-485 network, the logical state of the bus is determined by the polarity relationship between the (V+) and (V-) signals

RS-485 Biasing Resistors & Termination Resistors   6.5.6.4

Long RS-485 networks with high BAUD rates may require 120 ohm terminating resistors at both ends of the network as seen in Figure 6i. Termination resistors reduce reflections which appear as ringing at the signal transitions points. The decision of whether to use termination resistors should be based on the cable length and the BAUD rate of the transmission. In most cases, BAUD rates under 19.2 K do not require termination resistors. If termination resistors are used, the network must be designed with the appropriate biasing resistors to ensure reliable communications.

Biasing resistors are responsible to ensure that a network remains in the idle state when all the network drivers are tri-stated (all transmitters turned 'off'). A 200mV differential must be present between (V+) and (V-) to keep the bus in an idle state. Most master devices include some high resistance biasing (50-100 kohms). When termination resistors are used, the resistance of the biasing resistors needs to be much lower to maintain the 200mV differential. Termination resistors also increase the dc loading of the network.

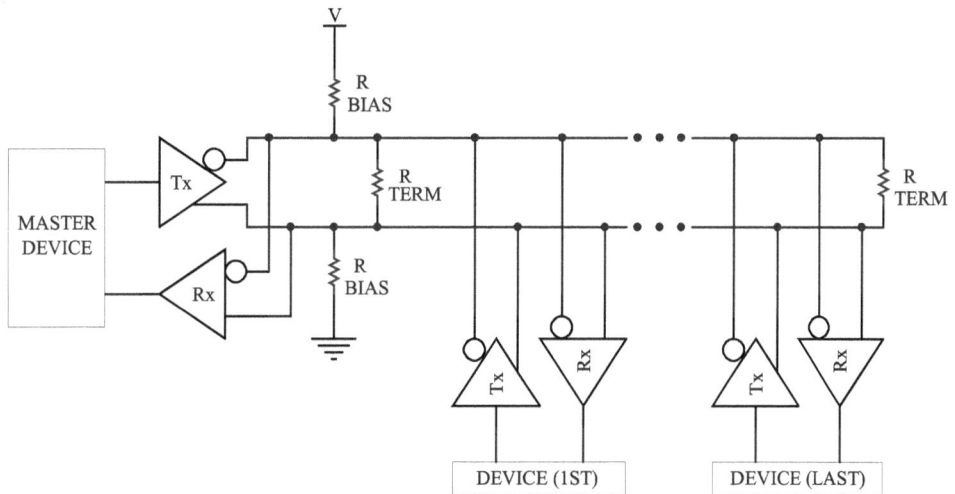

Figure 6i: Example where terminating resistors (R_{TERM}) are added to an RS-485 network, typically for a long network that uses BAUD rates above 19.2K, to reduce reflections. The network has biasing resistors (R_{BIAS}) as well to ensure that communications are reliable

Network design, including termination resistor and biasing resistor calculations depend on the type and number of devices on the network, the cable length, and the network BAUD rate. For more information on biasing and termination resistor calculations and recommendations, refer to:

<u>EIA/TIA-485 Standard</u>

- Telecommunications Industry Association
- http://www.tiaonline.org/standards/catalog/

<u>RS-422/RS-485 Application Note</u>

- Copyright: B&B Electronics
- http://www.bb-elec.com/tech_articles/rs422_485_app_note/table_of_contents.asp
- http://www.ATSeries.net/PDFs/RS422+485AppNote.pdf

Ethernet   6.5.6.5

Ethernet is a family of computer networking technologies commonly used in local area networks (LANs). Several variants of Ethernet are available. Older networks utilized coaxial cable to create the connections. Present networks utilize standard cables of multiple copper twisted-pairs with RJ45 connectors on each end. Speeds of 10, 100, and 1000 megabits per second are supported. An Ethernet network permits many devices to be interconnected and allows the devices to communicate via multiple protocols simultaneously.

Connecting a battery charger to an Ethernet network is as simple as plugging in the RJ45 cable. The connection to the network via hubs, switches, and gateways is no different than the connection for a desktop personal computer. The charger will require some configuration to interface to the network. The IP Address, the Netmask, and Gateway Address are standard configuration parameters utilized by all Ethernet devices. Security, access, and firewall configuration are well beyond the scope of this book.

The network topology is typically defined and specified by the Information Technology (IT) department and the network administrator.

Isolation   6.5.6.6

The charger communications discussed thus far have been wired connections which require the charger to be electrically connected to the network. The network may consist of many devices spread out over thousands of feet, meters, or miles. Without isolation, the charger's electronics could be damaged by currents and surges on the network. It is smart, therefore, to isolate the communications circuit from the rest of the charger's electronics.

TECH TIP

The isolation barrier should be galvanic such that no current flows between the communications electronics and the rest of the charger. Each side of the isolation barrier will have a separate ground and power supply. There is no direct conduction path across the isolation barrier; the information or data is passed from one side to the other magnetically or optically.

Ethernet driver and receiver circuits use transformers to magnetically isolate the charger electronics from the network cable connection. Many Ethernet RJ45 sockets have the isolation magnetics built into the connector shell, as pictured in Figure 6j.

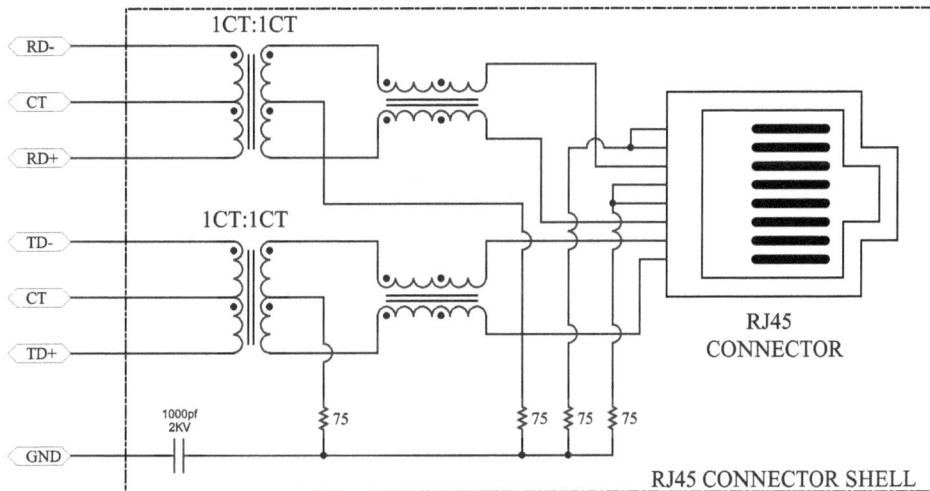

Figure 6j: An Ethernet network that isolates charger electronics from the network cable using transformers. Image courtesy of Pulse Electronics; available directly at https://www.networking.pulseelectronics.com/.

Optical isolation is typically used on RS-232 and RS-485 ports. Optocouplers use light to communicate across the isolation barrier. An optocoupler is a device that has an emitting LED and a detector circuit inside the same semiconductor device. The following example (Figure 6k) shows an RS-485 transceiver isolated from the core circuitry using three optocouplers. Notice each side of the isolation barrier has a different power supply.

Fiber Optic Interface 6.5.6.7

The networks described so far rely on copper wires to interconnect the network devices. The last section discussed isolation techniques to protect the charger electronics from the network bus, but the communication between the devices still rely on electrical pulses sent through a copper wire.

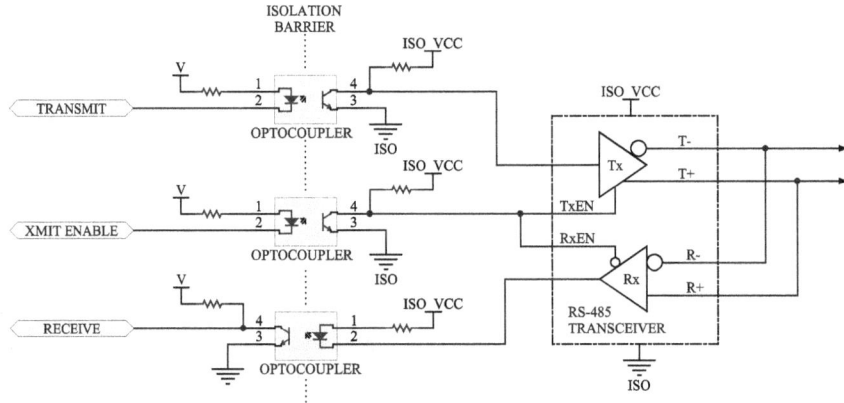

Figure 6k: An RS-485 network that is isolated using three optocouplers

Fiber optic networks communicate through either glass or plastic fibers. Each fiber has a light source on one end and a light detector on the other. Depending on the bandwidth and fiber type, cable lengths over a mile can be supported without any repeaters. Since no copper connections are required, all network devices are electrically isolated from each other. When a charger is connected to a fiber optic network, it too is isolated from other network devices; this protects its electronics from network surges.

Fiber modems are available from several manufacturers. These devices convert standard RS-232 and/or RS-485 signals to a serial fiber transceiver. Any RS-232 or RS-485 network connection can be converted to a fiber interface by replacing the copper network with fiber optic cables and inserting one of these fiber modems at each end of the fibers.

Fiber translators or media converters are available for Ethernet applications as well. There are numerous fiber formats. These vary by connector type, fiber type, and wavelengths and BAUD rates supported. Verify the parameters and details of the fiber network before specifying a fiber interface for your charger.

PROTOCOLS   6.5.7

The first sections of this chapter covered electrical standards and configuration options. Once an electrical standard has been selected and the charger is wired to the network, the configuration parameters need to be set correctly. Next, a network connection can be established and the transferring of data can begin.

The protocol specifies how to interpret the data being transferred between the devices on the network. Think of the network connecting the devices like a phone line connecting two people. The protocol between the two devices would represent the language being spoken by the two people. If the two people aren't speaking the same language, their voices (the data) will get transferred through the phone, but neither person will know how to interpret the data.

Protocols are like languages; they are rules for how data is interpreted between two devices. Languages use words to communicate; protocols use numbers. Protocols in their simplest form outline a few command formats, assign the commands to numbers, and define the rules required to format the data into the commands that have been established.

DATA TYPES, REGISTER MAPS & MINIMAL COMMAND SET   6.5.8

At a minimum, two types of data need to be represented: binary (bits) and analog. A binary is used to specify a parameter such as an alarm that only has two states. The alarm is either ACTIVE or NOT ACTIVE. Examples of the status of other binary data would be ON/OFF, OPEN/CLOSE, ENABLE/DISABLE, or HIGH/LOW. Analog data specifies a parameter with a value such as a measurement. Voltage, current, temperature, and alarm set points are all examples of analog parameters.

The format of the data is specified by the protocol as well. Bytes are the basic building block of serial communications and most protocols size the data types to fit into one or more bytes. Eight binaries (or bits) are packed into one byte. Analogs are one, two or four bytes in length. The format of the data and packing order of the data are all specified by the protocol rules.

Based on the rules of the protocol, a register map will be created by the charger manufacturer. The register map will list the relevant parameters grouped by type. Most protocols will distinguish between analog values that can only be read and analog values that can be read or changed. A measured value such as the dc bus voltage can only be read by an external device. A changeable set point such as an Equalize Timer can be read or changed by an external device. The same is true with binary data types. The status of an alarm can only be read. A control binary, such as FLOAT/EQUALIZE control, can be read to determine the present charge mode status or written to change the charge mode.

MODBUS   6.5.9

The Modbus protocol was created in the late 1970s by Modicon, a Programmable Logic Controller (PLC) manufacturer. The protocol is simple, royalty free and has become a standard for networking electrical industrial equipment. Modbus is a master/slave protocol such that one master device on the network is responsible for polling the remaining slave devices. The slave devices only respond when they are polled by the master. Chargers typically are Modbus slave devices and get polled by Modbus master devices such as Remote Terminal Units (RTU) and Programmable Logic Controllers (PLC).

There are two versions of the protocol: Modbus-ASCII and Modbus-RTU. The ASCII version of Modbus communicates via the seven-bit ASCII code character set. The messages are delineated by special ASCII characters. Modbus-ASCII was designed for older slower equipment that could not meet the timing requirements of the Modbus-RTU version. Modbus-ASCII is a legacy protocol and is rarely used.

Modbus-RTU utilizes eight-bit hexadecimal numbers. Messages are delineated by a 3.5-character timeout before and after each message. This character gap is used to synchronize all devices connected to the network. This character gap requires all frames within the message to have less than a 3.5-character gap or the frames will be interpreted as being in different messages.

The Charger Register Map would include registers for:

- Coils: for readable and writable binaries
- Binary Inputs: for read-only binaries
- Holding Registers: for readable and writable analogs
- Input Registers: for read-only analogs

The suggested minimal command set would include:

- Function Code 1: Read Coils
- Function Code 2: Read Binary Inputs
- Function Code 3: Read Holding Registers
- Function Code 4: Read Input Registers
- Function Code 5: Force (Write to) a Coil
- Function Code 6: Write Holding Register

All Modbus devices must be assigned a Modbus address. Modbus master devices direct commands to slave devices based on the slave's Modbus address. The format of all Modbus poll-request commands is similar. The commands all start with the Modbus address of the device the poll is directed to and the function code to execute. After the function code, an address field follows to specify which registers are to be operated on. For read commands, these will be the registers' values returned to the master. For write commands, these will be the registers to be written to, with the data following the register address.

Both the ASCII and RTU version of Modbus include a checksum at the end of each message. The checksums types are calculated differently. Modbus-ASCII uses an LRC; Modbus-RTU uses a CRC. Before a message is processed by a master or slave, a checksum is run on the received bytes of the message. If the checksum doesn't match the checksum at the end of the message, the message is considered invalid and is not parsed.

For more details about the Modbus protocol visit www.Modbus.org. The specification can be downloaded from the site. The specification will include more information on the data types and commands supported, how to calculate the LRCs and CRCs, exception codes, and much more.

DNP3   6.5.10

DNP stands for Distributed Network Protocol. As described at www.dnp.org, "The development of DNP3 was a comprehensive effort to achieve open, standards-based interoperability between substation outstations and master stations for the electric utility industry. Since its inception[4], DNP3 has also become widely utilized in adjacent

---

4 The DNP3 protocol was created in 1993 by GE Harris.

industries such as water/waste water, transportation and the oil and gas industry." In fact, DNP3 is based on IEC 60870-5 which was envisioned to become the standard protocol utilized by electrical utilities at the substation level.

The DNP3 protocol is much more complex than Modbus. DNP3 can be implemented as a master/slave protocol for basic applications. The protocol also supports multiple masters and report-by-exception modes such that one master device doesn't control the network access. In addition, DNP3 has many more data types and variations of data types such as frozen values, changed values (with and without time), and values with and without flags.

There are several subset levels of DNP3 defined to ensure DNP3 operability between simple lower level devices. The lower subset levels only need to parse a few commands and data types. Subset Level I would satisfy most charger communication requirements. The DNP3 protocol refers to data items as points, not registers.

The Charger DNP3 Point map would include points for the following data objects:

- Binary Inputs: for read-only binaries
- Binary Outputs: for readable and writable binaries
- Analog Input: for read-only analogs
- Analog Outputs: for readable and writable analogs

The suggested minimal command set would include:

- Class 0 Read: This would return all points
- Control Relay Output Block: This would implement writes to binaries
- Direct Operate Analog Output Block: This would implement writes to analogs

All DNP3 devices must be assigned a DNP3 address. All DNP3 messages have a header that specifies the source and destination of the message. The simplest way to poll data from a simple DNP3 slave device is to issue a CLASS 0 Read command. The device will respond with all the data points in the point map.

The DNP3 master device will need to issue a Control Relay Output Block (CROB) to change the state of any of the Binary Outputs (control bits). A Direct Operate command must be sent to change the state of any of the Analog Outputs. All DNP3 commands have multiple CRC checksums interspersed throughout each message. If a checksum doesn't match, the message is considered invalid and is not parsed.

For more details about the DNP3 protocol visit: www.dnp.org.

IEC 61850   6.5.11

IEC 61850 is an international standard defining the communication protocols for devices in electrical substations. This standard was developed by an IEC project group consisting of members from various vendors and countries. The goal of the group was to define abstract data models that can be mapped to several protocols. IEC 61850 devices use Ethernet TCP/IP networks to communicate with each other. Presently the mappings support MMS (Manufacturing Message Specification), GOOSE (Generic Object-Oriented Substation Event) and SMV (Sampled Measurement Values).

As mentioned in the previous section, DNP3 was developed and adopted to simplify equipment interoperability within the substation. Equipment manufacturers standardized on the DNP3 protocol to eliminate the need for the utilities to support multiple protocols and purchase protocol translators. All the substation devices could now speak the same language.

IEC 61850 takes this standardization to the next level; it standardizes the data naming and mapping as well. IEC 61850 is an object-oriented protocol with a standardized naming format. Devices that support DNP3 have tables of points grouped by their data type (binary, analog, input, output, etc.). Each DNP3 device has its own unique data maps and point names. A writable data value such as "High DC Voltage Alarm" is common for all chargers and charger manufacturers. Most chargers will have this same data point, but this data point will have a different name, point number, and location in each manufacturer's charger.

IEC 61850 eliminates the need for the utility to review the manuals of each device in the substation to determine how to request the specific data values (or data points) they require. IEC 61850 standardizes the names of the data values. The standard names are more readable and intuitive since they contain characters and numbers.

This section was meant to provide a basic understanding of the differences between IEC 61850 and DNP3. The details of the IEC 61850 protocol are well beyond the scope of this text. At the time of printing, Modbus and DNP3 are the most commonly used protocols supported by utility-industry stationary battery chargers. Substation devices that support IEC 61850 are becoming more commonly available. The assumption is eventually the demand for chargers with IEC 61850 support will also grow.

UNCOMMON (BUT USEFUL) OPTIONS FOR SCR/ SCRF CHARGERS

FRONT PANEL METER OPTIONS   6.6.1

1% Analog Meters   6.6.1.1

The standard front panel dc meters (1 voltmeter, 1 ammeter) are 3½ inch analog meters with 2% accuracy. Accuracy is rated at the full-scale deflection of the meter. So, for example, in a 130 Vdc charger, with a 200 V meter, the reading at any point has an uncertainty of ±2 volts.

You can improve on that by ordering 1% accuracy meters. The meters look identical, but now reading uncertainty is ±1 volt.

In either case, the displayed voltage isn't accurate enough for adjusting the float or equalize voltage. As noted in the charger instruction manuals, you should use a portable digital meter with at least 0.25% accuracy.

TECH TIP

Switchboard Meters   6.6.1.2

Or, you could order switchboard meters. These are optional 4.5 inch analog meters with a 240 degree dial and 1% accuracy. Because of their larger size, they aren't available for chargers in the Style 1A enclosure (refer to the SCR/SCRF sales brochure to see what charger ratings are affected).

The dial is round, although the front bezel of the meter is still square. To use switchboard meters, you need a lot more panel space, but you pick up a longer dial, which makes it easier to eyeball the voltage or current within 1%. But even with that accuracy and improved resolution, you still need a portable digital multimeter to calibrate float and equalize voltages properly.

Digital Meters   6.6.1.3

Or, you could order digital front panel meters. This option provides 3½ digit LED meters for both the dc voltmeter and dc ammeter. The display is a red LED, 0.56 inch high. Sorry, you can't order one digital and one analog meter. Also, digital meters aren't available for the small chargers normally supplied in Style 1A enclosures.

The option specifies an accuracy of ±0.1%, so for the 130 Vdc charger, where the meter has a full-scale rating of 200V, the reading is within 0.2 V. For the ammeter, the full-scale reading depends on the current shunt used in the option. One example: For a 100 A charger, a 150 A/50 mV shunt is used, so the accuracy of the meter display is within 0.15 A.

But as with any digital meter (including your portable meter), the last digit is always fuzzy. Allow another ±1 digit for any reading. Practically, this means that the ammeter in the 100 A charger, given in the example above, really is accurate within 0.2 A instead of 0.15A.

Digital meters are normally supplied for connection to 120 Vac power, which comes from the charger's main ac power input. This means that the meters go dark during an ac power failure, a difficulty not shared by analog meters. You can provide your own 120 Vac from a backup source, however, so that meter operation is continuous.

AC Voltmeter & AC Ammeter   6.6.1.4

These meter options, available separately, indicate the primary ac voltage supplied to the charger and the line current to the charger. For three-phase chargers, a phase or line selector switch is provided so that the user can read each phase individually. The standard meters are 3½ inch analog meters, with the same "look and feel" as the standard dc meters. As is true for the digital meters, the options aren't available for chargers in Style 1A enclosures because of the lack of usable panel space.

The ac meters are of the moving-iron type; they display the true rms value of the voltage or current with 2% accuracy. For chargers with ac input current over about 35 Aac, the ammeter uses a current transformer; three-phase chargers have three current transformers.

These options may be useful for sites that are cramped for space, since you can put ac instrumentation right into the charger. But the ammeter is of limited value in a standby application because the ac input current would spend most of its life near zero. You don't get much useful information from a 50 Aac meter, with an accuracy of 2%, when it's indicating only 1 Aac.

Remote Meters   6.6.1.5

Sure, you can have remote meters. This option doesn't actually supply the meter(s), just the connection points in an SCR/SCRF charger. You can connect a remote dc voltmeter or remote dc ammeter to the terminals provided (optionally) on TB3. You want the meter, too? Just ask.

Just remember, please, that the wiring to a remote voltmeter carries the battery (or charger output) voltage, and the wiring to a remote ammeter carries the shunt signal that feeds the charger's front panel meter and controls current limit. If you short-circuit either one of these signals, there will be big trouble. We recommend that you install a 1 A, 250 Vdc fuse in series with the remote voltmeter.

WHAT ARE MY OPTIONS FOR DETECTING BATTERY DISCHARGE?   6.6.2

Battery discharge occurs, of course, during an ac power failure, but may also take place if the instantaneous dc load exceeds the capability of the charger – not an unusual occurrence, for example, when starting a pump motor at a site.

You have two options: You can add a front panel ammeter to display battery discharge current, or an alarm circuit to send a contact closure when discharge occurs. In other words, in the spirit of Heisenberg, you can see it or send it, but not both.

To answer your next question, yes, some users have ordered both, and Engineering will do it.

Zero Center Ammeter   6.6.2.1

The ammeter displays battery charge or discharge current. A separate current shunt is installed, and the battery and load wiring are arranged so that the current shunt measures only the battery current. Positive deflection of the meter (to the right in the Northern Hemisphere) indicates charge current; negative deflection indicates discharge.

The key to ordering a battery discharge ammeter is to specify the maximum discharge current that the battery will see. This is usually more, sometimes a lot more, than the charger output rating. If the ammeter option has the same rating as the charger, the meter movement will probably be pegged (deflected to maximum) during battery discharge, and there is a risk of damage to the current shunt.

Battery Discharge Alarm   6.6.2.2

The Battery Discharge Alarm uses the same current shunt as the discharge ammeter (and the alarm should be specified for the same maximum discharge current, as discussed above). The shunt signal drives an electronic detector that can be adjusted from 2% to about 50% of the shunt rating.

Example: If you know that your battery will see a 150 A discharge, order a 200 A discharge alarm. It can be adjusted to send an alarm (transfer the alarm relay contacts) on a discharge current as low as 4 Adc, or up to about 100 A. This adjustment can be made either at the factory or in the field.

COMMON ALARMS (SUMMARY ALARMS)   6.6.3

We briefly described the common alarm circuit in the CASM on PAGE 88. If you have one or more legacy alarms, such as those described in this section, you may want to group their relay contacts into a single alarm contact. Here's the menu:

Common Alarm Buzzer   6.6.3.1

The Common Alarm Buzzer provides an 80 dB(A) horn, about 2,900 Hz, to alert you that any one of the included alarms has been triggered. Fine for an attended site, but if no one's there to hear it....

The signal is loud enough and high enough to be heard above most industrial noise, without quite being at the level that would damage hearing. If you can't hear it, of course, then there is already too much noise at the site.

An optional lightning arrestor (secondary arrestor) is available that increases the surge current capability to at least 10,000 A peak for the 8 x 20 μsec waveform. This offers substantial extra protection for induced surges. If a user wants primary surge protection (that is, protection against a direct lightning strike), he must install lightning protection at the building service entrance or distribution panel for his location. But note that ANSI C62.41 allows the use of secondary arrestors at these locations.

Common Alarm Buzzer for CASM   6.6.3.2

If you have a CASM, and you don't want any other alarms, you can also get a common alarm buzzer for it. It doesn't take the place of the common alarm relay on the CASM; you still have that. The option uses the same buzzer as in the legacy common alarm. The option includes a toggle switch to disable the buzzer, which also acts as a "call-back" to remind you to re-enable the buzzer when the fault condition has been cleared.

Common Alarm Relay with Buzzer   6.6.3.3

If you have a group of legacy alarms, you can get a Common Alarm Relay either with or without a buzzer included. Incidentally, are we stretching the meaning by calling a 2,900 Hz tone a buzz?

The common alarm relay works just as it does in the CASM but includes any legacy alarms you may have ordered. It can also include the CASM, of course, by wiring the CASM common alarm in with the legacy common alarm. That's a lot of commonality.

CABINET HEATERS   6.6.4

Internal cabinet heaters are useful if the charger is to be stored for an extended period in an unheated location. The heaters provide just enough energy to keep frost from condensing on critical components, or for that matter, any components at all. Of course, you need to provide power for the heaters: 120 Vac or 240 Vac, 50 Hz or 60 Hz, up to about 450 W, depending on the cabinet size.

The option includes a thermostat control, set to 70 °F (about 20 °C), and a separate single-pole circuit breaker.

HOW CAN I PROTECT MY CHARGER FROM LIGHTNING? 6.6.5

Here is a quote from HindlePower's Charger Specs FAQ:

All chargers are protected by MOVs (metal oxide varistors), on the ac input and dc output, that meet the requirements of IEEE/ANSI C62.41 for Category B locations.

The IEEE standard defines test methods for protection devices, such as MOVs. The test waveform for induced surges in indoor locations, Category B, is:

6 kV open circuit, 50 ohm source impedance, 1.2 x 50 μsec unipolar

3 kA peak, 8 x 20 μsec unipolar

What the heck does all this mean? It means that the charger ac input and dc output are protected by MOVs that have been tested to withstand 3,000 A peak current from a lightning surge that lasts 20 microseconds. Once. Maybe several times. These are called induced surges. That is, they are a secondary effect – a transient caused by current flowing in power lines, or even the ground, due to a lightning strike elsewhere. A direct lightning strike can be much more damaging.

If you think you're in a location that needs more protection, the optional lightning arrestor can take a 10,000 A beating and keep on ticking. But note that the onus is on the user to ensure that the primary service entrance is protected against a direct lightning strike. So if you're in a location with frequent, severe thunderstorms, the lightning arrestor offers good secondary protection at the charger location, but make sure your service entrance protection is up to snuff.

TECH TIP

» *Q: How do I know if the MOVs in my charger are still OK?*

A: *The MOV failure mode is a helpful indication. After several surges, the withstand voltage gradually deteriorates, and eventually the MOV fails when a surge overwhelms it, and it vaporizes, usually with a loud report. It may leave its pigtail leads behind. If you find that an MOV is missing, it should be replaced as soon as possible. We don't recommend elective replacement of MOVs, but I've read that some advisors recommend that for consumer equipment.*

SWC FILTERS   6.6.6

Time for another excerpt from HindlePower's standard documentation:

Oscillatory surges (IEEE 472/ANSI C37.90 Surge Withstand Capability (SWC))

Test Waveform: 2.5 – 3.0 kV initial peak, ringing at 1.0 – 1.5 MHz,

decay to 50% in 60 µsec

Repetitive at 50 Hz minimum

Apply for 2 seconds minimum

NOTE: There are two levels of this requirement:

Level 1 (NEMA PE 5-1991) The equipment must survive the test waveform without
damage or any degradation in performance.

Level 2 (ANSI C37.90) The equipment must produce no erroneous indications during
the application of the test waveform. We interpret this to mean no erroneous alarms,
and no more than a 10% change in output current.

The SCRF Series meets Level 1 with no additional filters. The optional SWC filters must
be added to meet level 2.

The AT Series meets all requirements with no additional filters required.

Note that the AT series meets the SWC surge requirements with no added filters, but
it was tested to the 1991 version of NEMA PE 5. The 1996 version requires the equipment
to pass at Level 2.

The SWC filter option applies to the SCR/SCRF charger and provides the necessary
immunity for the original oscillatory surge waveform. ANSI C37.90 also includes a "fast
transient" test. The AT charger (but not the SCR/SCRF) meets this requirement.

OVER-TEMPERATURE ALARMS   6.6.7

These are covered in Chapter 5, *Temperature Effects*.

HIGH RIPPLE ALARM   6.6.8

You know that dc filters use electrolytic capacitors, and that, traditionally, they have required eventual replacement because of gradual loss of capacitance. This is the only charger component that we suggest you might consider electively replacing after ten years. First, though, there is nothing sacred about ten years. Capacitor manufacturers insist that their products are more durable than that. Second, we also say that the best indicator is to measure the output ripple voltage during routine maintenance, and if it starts increasing, it's time to act.

The High Ripple Alarm is an option that could help keep an eye, figuratively, on your filter capacitor life. If the output ripple out of the box is 100 mV, and you set the ripple alarm to trigger the alarm relay at 200 mV, you have a good chance of nipping any problem in the bud. You can set the alarm threshold higher, up to what you think is tolerable for your system. A level of 500 mV on a 130Vdc bus is probably good.

There's a cost associated with this, of course. You have to trade off the purchase price against the likelihood that you will use this only once or twice during the life of the charger, and factor in any degradation in system performance due to undetected high ripple voltage. You must also consider the ease of measuring ripple voltage manually during routine maintenance.

It seems obvious that you wouldn't order this option with an unfiltered charger. But think about this: Since ripple voltage on the battery is a function of battery impedance, a High Ripple Alarm might detect a rise in battery impedance. There are a lot of other factors to consider, though. Other loads on the dc bus (UPS, for example) can contribute ripple voltage to the bus, and affect alarm operation. The bottom line? High ripple is usually caused by the charger, but not always.

CURRENT LIMIT ALARM   6.6.9

We get quite a few requests for this option, and it mystifies me. First of all, you shouldn't be alarmed that the charger is in current limit. That's a normal operating mode. You should only worry if the charger *isn't* in current limit when it's *supposed* to be because then something is wrong.

It's good to remember that in a dc system, the battery sets the voltage. Not the charger. The battery. (Refer to *Battery Chargers: How Different from Power Supplies?* in Section 2.1.4.)

The charger feeds current to the battery, and the battery decides what voltage it's going to show to the world, based on the charge current and its own state of charge. The only time the charger thinks it has the upper hand is when the battery is fully charged, sitting at float voltage, and the charger has to provide only enough current to maintain the float voltage.

This means that any time the battery is *less than* fully charged, it will present a terminal voltage less than float voltage. When a charger sees this, it says, "OK, I have to send more current to try to bring that voltage up," and will wind the output current all the way up to current limit, if necessary.

So, the Current Limit Alarm is, in essence, a "battery isn't fully charged" indicator.

In the AT charger, a current limit signal is available through the communications option, using MODBUS or DNP3. In the SCR/SCRF charger, the alarm is an add-on circuit board that has to be calibrated to a specific current limit value. If you change the current limit setting in the field, you have to recalibrate the current limit alarm also. Is it worth it?

ARE THERE VENTILATION OPTIONS
TO PREVENT HYDROGEN BUILDUP?   6.6.10

SAFE PRACTICE

Flooded batteries are occasionally equalized. During equalization, some hydrogen is inevitably evolved, and it's a good idea to try to get rid of it before it turns your facility into a famous dirigible. Usually, with proper charging and proper site design, hydrogen will not build up to an explosive level.

But your site plan might call for forced ventilation of the battery room during equalize charging. An optional relay contact that closes on equalize can be provided to control an external contactor. You can also order a Ventilation Fan Control option, complete with a contactor rated for 10 Aac or 20 Aac at 240 Vac. Breathe easy.

COUNTER-EMF DEVICES   6.6.11

Back in SECTION 1.5.5 on *Equalization Charging*, we mentioned counter-EMF cells, used to standardize the voltage applied to loads on the dc bus during equalization. When a battery is equalized, the terminal voltage may rise to a level higher than the dc loads can tolerate. A counter-EMF cell can be inserted in series with the load to reduce the voltage to an acceptable level.

Today, counter-EMF devices are solid state, using silicon diodes. There are many decisions you need to make when specifying a device, depending on the load current level, the voltage to be dropped, and the control scheme, which may create multiple voltage drops.

Because there is heat to be dissipated, such devices take up a fair amount of room. But once installed, they need no maintenance or electrolyte replacement, as was historically required for electrolytic counter-EMF cells.

APPLICATIONS

MOST DC SYSTEMS, such as those in substations, are straightforward – one charger, one battery, and a dc load consisting of relays, circuit breakers and such. Unusual cases do come up, though. Here are a few words about how to handle special applications.

The author had access to examples and data from his former employer, HindlePower, but the principles apply equally when considering these applications while using any manufacturer's products.

CONNECTING CHARGERS IN PARALLEL

REDUNDANT & NON-REDUNDANT OPERATION 7.1.1

TECH TIP

Your site plan might call for having two chargers, connected to the battery in parallel, to provide redundancy. As you know by now, any two chargers with the same output voltage can be used in parallel, and both will provide load current as needed. The rule on redundancy is that each charger must be rated to supply the entire standing load, plus about 10% more to charge the battery when needed. That means that if you have two differently rated chargers, the smaller must stand by itself, in case the higher rated charger takes a vacation.

This applies even in the case where you have redundant batteries. If you have one charger per battery, treat the charger sizing for each battery as if it were a stand-alone system.

LOAD SHARING   7.1.2

Some sites call for load sharing. *Battery chargers don't need to share the dc load current to work successfully in parallel.* Some users require load sharing, though, because it allows them to confirm at a glance that both chargers are operating normally. It also prevents that annoying (and possibly disruptive) false zero-current alarm, which might occur if one charger isn't contributing load current. There is also a slight improvement in transient response in the event that one charger should fail, but with a battery on line, this isn't so important.

Check back at the description of the charger failure alarm in the CASM in Section 6.3.5. If you have an SCR/SCRF charger, with a CASM installed, and select the True CFA alarm mode, you will be immune from false zero current or charger failure alarms, even without load sharing. For more information on sharing the load between chargers see Section 6.4.2.

Note, though, that IEEE Standard 946 recommends that forced load sharing "should" be specified for parallel chargers.

STARTING DC MOTORS: HANDLING SUDDEN LOAD DEMANDS

A FREQUENT REQUIREMENT at a site is to start a dc motor, a lubrication pump, or whatever. When a motor is first started, its inrush current must be supplied by the dc bus. If the inrush current exceeds the charger rating (which is usually the case), then the battery must make up the difference. The sudden demand depresses the battery voltage slightly, ensuring that the charger enters current limit (for more on this behavior, see *Current Limit Alarm* in Section 6.6.9).

The SCR/SCRF chargers and the AT chargers respond differently to these sudden load demands. The current limit control on the AT is fast, usually acting within one cycle of a 60 Hz ac line. The SCR/SCRF current limit control doesn't take full effect for about 100 msec, or about six cycles.

The SCR/SCRF charger is normally supplied with a fast-acting fuse in the dc output circuit. When the dc motor is first connected across the dc bus, it looks almost like a short circuit. The delay in the current limit circuit can allow enough charger output current to flow (to the motor) to clear the fuse. To ensure reliable starting of dc motors with the SCR/SCRF charger, we recommend that you install the optional motor starting kit, which consists of a freewheeling diode to be wired across the output of the charger (but inside the dc output circuit breaker), and an upgrade to a slightly slower fuse.

» *Q:* *With a slower fuse, is the charger still adequately protected?*

A: *Surely. If the purpose of the fuse were to protect the dc bus from excessive charger output current, the motor-starting fuse would be up to the task, since it has the same rating as the standard dc fuse. Actually, the more important role of the dc fuse is to protect the dc bus from an internal fault in the charger. The motor-starting fuse will also do that.*

The AT charger is designed to withstand dc motor starting requirements with no modifications. More is said about motor starting in Section 4.4.1.

The following are two case studies to help illuminate what happens in dc motor starting. In the first, a customer kindly provided oscillographic test data for starting a large pump motor; their data covered the first 100 ms or so of motor starting. We were asked to analyze the data to find the fuse rating that would withstand the motor inrush current.

CASE STUDY 1: 7.2.1

In Figure 7a, the horizontal axis ranges from 0.0 to 1.1 seconds. The vertical axis shows dc volts.

Based on the motor inrush and charger output current data provided by the user, we developed a simple equivalent circuit for the battery-charger-motor installation and built an incremental model. We used the model to analyze the behavior of the circuit to estimate the wiring resistances, and predict the behavior of the battery charger, assuming that the dc fuse wouldn't blow.

Assumptions:

- The battery voltage decreases to the open-circuit voltage as soon as the motor load is applied.
- The battery open-circuit voltage is 123 V.
- The battery charger's bridge-output voltage is 132 V, and doesn't change for the first 80 msec after the load is applied.

- At 80 msec, the motor current is 640 A. The charger supplies 280 A of this and the battery supplies the remaining 360 A. This comes from the user's data.
- The battery internal impedance is 0.015 Ω.
- Reverse current cannot flow through the charger. Also, after 200 msec, the charger output is limited to 83 A. This has been modeled as a geometric decrease in current, converging to 83 A.
- V(m), the motor's back EMF, is described by V(m)= B × t, where B is empirically derived. V(m) becomes constant at 0.6 seconds.
- The motor armature resistance is 0.1 Ω, based on 5% per unit impedance at 60 A full load.
- The armature self-inductance is 6 mH.
- The battery charger's dc circuit inductance is 4 mH, but the model works better if we assume an average inductance of 3 mH, since the inductance decreases at high currents.

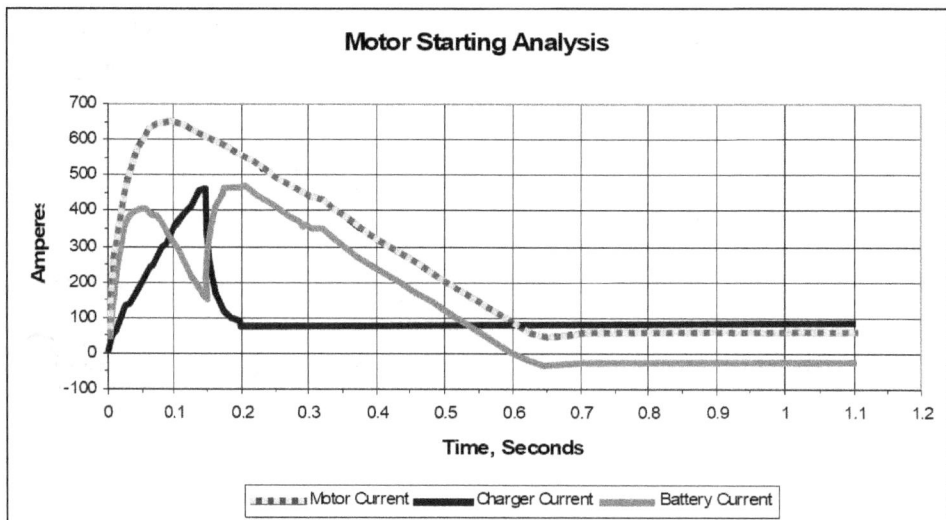

Figure 7a: Model of current over time for a circuit containing a battery connected to a charger that must start a motor. Based on customer-supplied oscilloscope data during the first 0.1 seconds of the graph

With these assumptions, the circuit was analyzed incrementally, resulting in the waveforms in Figure 7a. This graph is hypothetical; it predicts the behavior of the charger current and the battery current if the charger dc fuse doesn't blow.

This model conforms closely to the actual behavior of the motor inrush current and the charger current during the first 80 milliseconds. The rest of the behavior is extrapolated to conform to the actual measured motor inrush current.

There is an unknown factor in the model: I assumed that the charger dc inductors would not saturate during this period. If they would saturate, this might change the portion of the inrush current provided by the charger.

Based on the predicted peak charger output current of 400 A, using a type JJN-125 fuse[1] for F1 in the charger will allow the motor to start without blowing the fuse.

The second case study is a field report of an erroneous charger failure alarm, due to the action of the current crowbar circuit during motor starting. The current crowbar is a safety feature to protect the charger against a sudden fault on the dc bus. When the current crowbar is activated, the charger is forced to return to soft start. Remember that starting a motor can look almost like a short circuit until the motor gets going.

CASE STUDY 2:   7.2.2

The user's original problem report for an AT charger rated at 24 V, 100 Adc was similar to the following:

EXAMPLE

Right now we are working on a problem with the fault signal from the charger. We have the charger connected to two 12 V, 100 Ah...batteries in series, going to a 24 V, 22 amp dc lube oil pump. When we first turn on our pump we get a low dc voltage alarm then a dc output failure.

After a minute, the output current [and] voltage stabilize and the faults go away. The output current from the charger starts out low, about 1 amp, and within a minute it stabilizes at 23 amps.

The slightly abridged analysis was (...the envelope, please...):

The user is trying to start a dc motor rated at 22 A. The inrush will be at least 220 A. That's typical: The inrush range varies from eight times to over 20 times the full load current, depending on the motor design.

---

1   JJN is the manufacturer's type designation.

The charger, of course, is limited to 110 A. The response time of the current limit circuit is fast, though not instantaneous. However, the response of the current crowbar is instantaneous. When the motor is started, the current demand is going to be shared by the battery and charger. The battery, surprisingly, can't deliver the necessary current instantaneously, and so the terminal voltage drops, possibly to less than 2.0 VPC (volts per cell).

This leaves the charger to try to make up the difference. The instantaneous current demand may activate the current crowbar, which sends the control circuit back to the beginning of a soft start. Thus, the dc bus voltage collapses by several volts, and will be determined by the characteristics of the battery alone for at least the next four seconds.

By that time, of course, the motor is up to speed, and the bus voltage has recovered to 2.0 VPC or higher. The battery is supplying the motor current, and the charger gives a dc output failure alarm because its voltage is below float, and its output current is below current limit – exactly the way we designed it. Once the charger output recovers to the float voltage, it begins to supply the motor current, and the alarms are reset.

I think, however, that the common alarm relay shouldn't transfer for 30 seconds, which should at least prevent any remote annunciator from being activated.

There are guidelines for calculating charger ratings based on load demands, battery size, recharge time, etc. Sales can help you with these. Unfortunately, these guidelines are inadequate when transient loads such as dc motors are thrown into the mix. You would think that rating the charger at five times the motor rating would be sufficient; in fact, it would be if it were an SCR charger. The issue here is "false" alarms (meaning a transient alarm that the user doesn't want to see).

Finally, the way a dc system responds to a motor starting event depends to some extent on the way it's wired. If the motor is electrically closer to the charger than the battery, you are more likely to see the behavior that the user describes.

FUSE COORDINATION   7.2.3

In the SCR/SCRF case study, you saw that the result was the choice of a specific fuse. For the purposes of motor starting, fuses are chosen to allow the peak charger-output current prior to the initiation of current limit, but still offer short-circuit protection for the rectifier SCRs and diodes.

CHARGERS IN SERIES: DO YOU REALLY WANT TO DO THIS?

THERE ARE MANY installations where two or more chargers may be connected in parallel. What about connecting chargers in series?

When you put two chargers in parallel, you know that the main requirement is that they have the same voltage ratings. They may have different current ratings if load current sharing isn't important.

But when you want to connect chargers in series, you must be more careful. You may see this situation at a site with two 130 Vdc batteries connected in series to make a 260 Vdc battery.

» *Q: Why isn't this just considered a 260 V battery? We should use a 260 V charger.*

    A: The reason for the 2 × 130 V designation is usually that the battery is tapped in the center, to provide power for both 130 Vdc and 260 Vdc loads. In this arrangement, the discharge rates on the two 130 V sections will be different. If you tried to charge the two sections in series, using a 260 V charger, one of the 130 V battery sections could be seriously overcharged. Or one could be seriously undercharged. So, users employ a separate 130 V charger for each individual section.

» *Q: That sounds OK to me. Why the big deal?*

    A: It is OK, until you start to add options to the chargers, like ground fault detection. When you do that, you're creating an ohmic connection between two chargers that should be isolated from each other because of their common connection to the battery midpoint.

Other problems can arise if someone wants to use two existing 130 V chargers to charge an existing 260 V battery that isn't tapped in the center. Now you really have to be careful: The current ratings must be identical, and the current limit settings on both chargers have to be the same. Even with careful settings, there could be problems in current limit with voltage sharing – the ability of the two chargers to share the battery voltage equally. Equal voltage sharing is crucial to avoid overstressing the chargers' internal components.

The bottom line: *you need, I repeat need, to talk to Engineering* before you put two chargers in series.

TECH TIP

MIXING CHARGERS ON A DC BUS (IN PARALLEL, OF COURSE)

WITH DIFFERENT CURRENT RATINGS?   7.4.1

You know by now that two chargers with the same voltage rating can always be connected in parallel. They may or may not share load current. If they have different current ratings, they certainly won't share the current equally.

In SCR/SCRF chargers, you can have proportional load sharing using the forced load sharing option. In the AT charger, it's OK to have parallel chargers with different current ratings, but you can't use load sharing at all. The AT control circuit simply won't allow it.

USING DIFFERENT CHARGER MODELS?   7.4.2

Users sometimes ask "Can I put an AT charger in parallel with an SCR/SCRF charger?" Yes, if they're the same voltage rating. But they won't get anywhere near load sharing. There is no way to force load sharing between different charger models.

FROM DIFFERENT MANUFACTURERS?   7.4.3

We knew this would come up. You want to use brand L on the same bus with an AT? We don't see any reason it won't work, but frankly, we haven't tested it. The only foreseeable problem is that changes in load current demand might cause some instability due to differences in transient response between the two chargers, and/or differences in the charger output transfer characteristics.

CAN I RUN A CHARGER FROM AN AUXILIARY GENERATOR?

A BATTERY CHARGER can be run from a diesel or gas turbine generator that is sized appropriately. You should be aware of performance issues that may affect other equipment running at the same time on the generator. What follows will help you to size a generator successfully.

SIZING THE GENERATOR   7.5.1

Handling the ac inrush current to the battery charger is an important consideration in sizing. The inrush current to a charger can be as high as 15 times the full load operating current, with a typical value of 10 times. When operating from a normal utility source, this magnitude lasts for only one cycle at the line frequency, about 16.7 ms at 60 Hz, and is usually greater in the first half cycle than in the second half cycle. This creates a dc component in the transformer core flux that must be overcome before the steady state is reached (the 15× inrush current specification takes the dc component into account).

The inrush current over subsequent cycles is much lower, and the current rapidly approaches the no-load value, usually within about ten cycles at the line frequency. The no-load value for most transformers is about two to four percent of the full load current. The behavior of a charger is similar, in this respect, to other ac machines, such as a compressor or pump motor.

When you run a charger from a generator, the inrush current is a more complicated issue. The very high, initial, peak inrush-current can easily be provided by a normal utility connection, whose short-circuit current capability can be 22 kA or greater. Even with this capacity during an inrush, the ac voltage at the input terminals of the charger will be depressed by an unpredictable amount, due to the impedances of the utility source, the facility wiring, and the charger's internal wiring. A good estimate is a 10% reduction in ac voltage at the input terminals for the duration of the inrush.

When a charger is started with a generator, the generator initially sees what looks like a short circuit on its output. A diesel generator might have a much lower fault-current capability than a normal utility, due to the finite reactance of the generator. One manufacturer specifies the *sub-transient reactance* as that which operates over the first six cycles after a short circuit is encountered. For their products, a reactance as high as 0.17 per unit is possible, but their example calculations use 0.10, which is a more conservative number for circuit breaker coordination.

EXAMPLE

Example: Suppose you are installing a 130 V, 300 A charger to run from a generator. The charger's normal full-load input current is 81 Aac at 480 Vac. If you obtain a 100 kVA generator at 480 Vac three phase, the generator's normal full-load output current would be approximately 120 Aac. With a 0.10 per unit reactance, the initial fault current for a short circuit will be about 1,200 Aac. With a full-load rating of 81 Aac, the charger would normally demand an inrush of between 800 Aac and 1,200 Aac. It is important to note, however, that for the first cycle or two, the output voltage of the generator will be near zero because its internal impedance is much higher than that of a normal utility connection. This changes the start-up dynamics: The charger transformer's inrush current will be lower and will take longer to reach steady state. Settling time cannot be predicted, only measured, and it will vary.

TECH TIP

Because of these considerations, we usually recommend that a generator be sized at twice the total input VA rating of the charger. In this example, this would mean a rating of about 150 kVA. We make this recommendation because we expect that other ac loads, besides the charger, will eventually be connected to the generator. If the generator is permanently dedicated to the charger, a lower rating is permissible, and probably is economically more desirable. A generator rating of at least 125% of the charger input VA rating would be satisfactory.

WHEN BAD THINGS HAPPEN

LIFE CAN BE interesting. Unfortunately, it isn't always nice. Sometimes, despite our best efforts to prevent trouble, we wind up wading into it deeply. Below you will find descriptions of what to expect when battery chargers experience extreme scenarios. Again, we've provided examples from our experience with HindlePower chargers, but the principles can be applied to all manufacturer's products.

WHAT HAPPENS IF A BUS FAULT OCCURS?

WHAT HAPPENS WHEN a battery charger is suddenly presented with a short circuit on its output terminals? There are a few possibilities, and they may be different for the SCR/SCRF and AT chargers. IEEE standard 1375-1998, "Guide for the Protection of Stationary Battery Systems," gives general guidelines for the charger contribution to a bus fault. Here are some specifics.

SCR/SCRF CHARGERS 8.1.1

If a charger has a dc filter, and especially if it has a battery eliminator filter, the dc output fuse almost certainly will clear on a "high quality" fault[1] , isolating the charger almost immediately from the dc bus.

The prospective current from the filter's output capacitor(s) can be very high – 10,000 A or higher. The charger's dc output fuse, however, is a current-limiting type, and the actual short-circuit current from the charger into the fault would be limited to 10,000 A. The fuse starts to clear in under one millisecond.

---

1   A "high quality" fault is one that will support little or no voltage on the dc bus.

What if the bus fault is "low quality," that is, it supports a few volts? In factory tests on a 24 V, 300 A charger, operating into a fault that supported about 3 Vdc, the dc fuse didn't clear. The charger delivered 2,000 A for 100 milliseconds, and then reduced its output to its current limit value of 330 Adc.

AT CHARGERS   8.1.2

The AT charger is designed to withstand dc bus faults without clearing output protection, whether circuit breaker or fuse. If the charger is filtered, the filter discharges into the fault, providing several hundred to a few thousand amperes, with a time constant determined by the impedances of the filter and charger wiring. Typically, that's less than a millisecond.

Fuse and circuit breaker manufacturers provide I^2t ratings, which are a measure of the energy that the fuse is able to handle without clearing. They can be used to determine the maximum cumulative (current $\times$ time) that a fuse will allow to pass. If you don't want the dc output protection to clear, choose a fuse (or breaker) with an I^2t rating greater than the prospective fault current when the output filter is discharged (usually less than 1 millisecond). We recommend using the standard dc circuit breaker, which is rated to withstand the fault current.

Figure 8a shows the prospective fault current from an AT10.1, 130 V, 50 A charger with a battery eliminator filter, assuming that the only limits to the current are the filter and charger wiring resistances.

After the filter is discharged, the AT charger contributes its current limit value to the fault.

WHAT HAPPENS IF A LOAD DUMP OCCURS?

A LOAD DUMP is a sudden large decrease in the dc output current of a battery charger. It may happen routinely, due to normal load control at a site, or abnormally, due to an emergency, system failure, or manual intervention.

If the disconnected load is inductive (a pump motor, for example), there could be a high voltage impressed on the dc bus. Normally, the battery can absorb the inductive energy, limiting the voltage rise. If the battery Ah rating is small, however, the voltage rise on the battery terminals could be substantial. This is a rare occurrence.

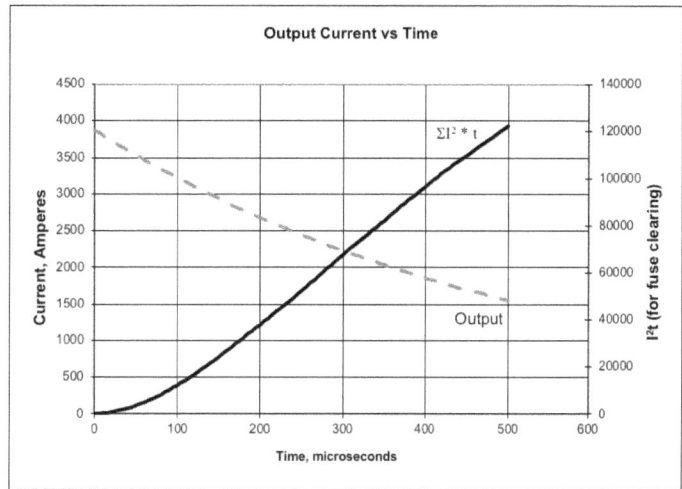

Output Current vs Time

Figure 8a: Prospective fault current from an AT10.1, 130 V, 50 A charger with a battery eliminator filter, assuming that the only limits to the current are the filter and charger wiring resistances. The rising graph shows the I²t rating, a measure of the energy that the fuse is able to handle without clearing

Another possible source of a voltage rise is the charger itself. A rectifier SCR, if it's conducting at the instant the load is dumped, continues to conduct until the end of the half cycle at the line frequency. The energy that it passes through to the output charges the filter capacitors, which may rise by several volts (again, unless the battery is able to absorb the excess energy).

The voltage rise due to inductive energy can be handled by a free-wheeling diode connected across the rectifier bridge. This is a standard component in the AT charger, and is included in the optional motor starting kit for the SCR/SCRF charger.

IS THE YEAR 2030 GOING TO CAUSE COMPUTER PROBLEMS?

YOU REMEMBER Y2K, of course. The year 2000, at 12:01 AM on January 1, when power generating stations were supposed to shut down, airplanes were going to fall out of the sky, and worst of all, your bank was going to forget your credit card number. Not to mention the year you were born. All due to computer programming limitations that were decades old.

Well, the world somehow muddled through it. During 1999, we were asked by many customers to verify that our battery chargers wouldn't come to a grinding halt on the first day of January 2000. This was relatively easy, since our power systems products didn't contain any electronic calendars, or even real-time clocks, and weren't connected to the internet or any net at all, except for wired-in SCADA systems.

Then, January came, and not much happened. This was probably due, at least in part, to the efforts of many programmers literally brought out of retirement to revise forty-year old code, which had scrimped on computer memory by storing only the last two digits of the year (that is, 60, 61, and so forth). So, in the legacy code, the years 2000, 2001, and so forth were indistinguishable from 1900, 1901, and so forth. [This was done primarily because data storage was expensive in 1960 – well over $1 per byte.]

Have you heard that this problem will arise again in 2030? Some believe it will because some programmers patched the code using a "windowing" method; they made programs think that the year '30 means 2030, while '31 means 1931 (the actual decade they used may vary). Of course, these programmers justified their choices by asserting, "These programs won't be around in 30 years," the same assertion they made in 1960.

Rest easy. HindlePower products will not be affected by the year 2030, or any other year.

STANDARDS & CODES

THE FOLLOWING ARE brief descriptions of industry standards, specifications, and codes that may influence the design or application of battery chargers and other equipment in a dc power system. Using chargers from one manufacturer, HindlePower, we show you which standards, specifications and codes apply; you may use this information when considering purchase of those or any competing chargers.

SIZING BATTERIES

THE IEEE PUBLISHES several standards, called "Recommended Practices," for sizing, installing, and maintaining storage batteries in various scenarios.

- IEEE 450 "IEEE Recommended Practice for Maintenance, Testing, and Replacement of Vented Lead-Acid Batteries for Stationary Applications"
- IEEE 484 "IEEE Recommended Practice for Installation Design and Installation of Vented Lead-Acid Batteries for Stationary Applications"
- IEEE 485 "IEEE Recommended Practice for Sizing Lead-Acid Batteries for Stationary Applications"
- IEEE 1013 "IEEE Recommended Practice for Sizing Lead-Acid Batteries for Stand-Alone Photovoltaic (PV) Systems"
- IEEE 1106 "IEEE Recommended Practice for Installation, Maintenance, Testing, and Replacement of Vented Nickel-Cadmium Batteries for Stationary Applications"
- IEEE 1115 "IEEE Recommended Practice for Sizing Nickel-Cadmium Batteries for Stationary Applications"
- IEEE 1184 "IEEE Guide for Batteries for Uninterruptible Power Supply Systems"
- IEEE 1187 "IEEE Recommended Practice for Installation Design and Installation of Valve-Regulated Lead-Acid Batteries for Stationary Applications"

SIZING CHARGERS

REFER TO IEEE Standard 946-2004, "IEEE Recommended Practice for the Design of DC Auxiliary Power Systems for Generating Systems." This standard was written originally for nuclear plants; in 1992, the scope was expanded to include all generating plants, since there were no new nuclear plants on the drawing boards.

PROTECTING BATTERIES

IEEE STANDARD 1375-1998 is the "IEEE Guide for the Protection of Stationary Battery Systems." There is a section that covers the contribution of a battery charger to a bus fault in a dc system.

INDUSTRIAL STANDARDS FOR CHARGER EQUIPMENT

NEMA STANDARDS   9.4.1

TECH TIP

PE 5, PV 5   9.4.1.1

NEMA Standard PE 5, "Utility Type Battery Chargers," is the principal specification covering the products discussed in this book. PV 5 is an older version, now retired.

PE 7, PV 7 (Communications Chargers)   9.4.1.2

Standard PE 7 covers chargers for communication-type batteries. The chargers discussed in this book meet most of the requirements of PE 7, but they haven't been tested to verify compliance with the EMI portions of the standard (note, however, that the AT series meets the EMI standards for CE conformity). PV 7 is an older version, now retired.

RI 2 (You're kidding, right?) 9.4.1.3

RI 2 is a really, really old version of the Utility Type Charger standard. Up until about 2000, it would still appear in customers' specifications, especially with regard to production testing, which was unnecessarily time-consuming. We're happy to see it disappear.

NEC (NFPA-70) 9.4.2

The NEC (National Electrical Code) covers a lot of territory. Our primary concern is with the requirements for wire sizing, both for internal charger wiring and for interfaces with field wiring.

SURGE PROTECTION 9.4.3

ANSI/IEEE C37.90, Surge Withstand Capability 9.4.3.1

This standard defines, among a lot of other things, the test requirements for oscillatory surge withstand capability. The standard is referenced also by the NEMA standard, PE 5.

ANSI C62.41, Guide on Surge Voltages 9.4.3.2

C62 sets standards for the testing and performance of surge suppression devices, which includes the MOV (metal oxide varistors) used for surge protection in the input and output circuits of HindlePower products.

SAFETY STANDARDS 9.4.4

UL 1012, UL 1564, CSA 22.2, & CE 9.4.4.1

The HindlePower AT series of industrial chargers is certified by CSA to CSA 22.2. They are also certified by CSA, as a NRTL (National Recognized Testing Laboratory), to meet the requirements of UL Standard 1012, "Power Units Other Than Class 2," and UL standard 1564, "Industrial Battery Chargers."

The AT series meets the requirements of the CE Low Voltage Directive, the EMC directive, and the pertinent EMC standards. HindlePower will provide a Declaration of Conformity for equipment that must meet CE requirements. This is an option that must be ordered separately.

CIRCUIT BREAKERS   9.4.5

UL 489, NEMA AB-1, & UL 508   9.4.5.1

All dc output circuit breakers are UL listed to UL 489 and meet the requirements of NEMA AB-1. DC circuit breakers rated for 100 A trip or lower have an ampere interrupting capacity (AIC) of 5,000 A. Breakers rated over 100 A trip have an AIC of 10,000 A or higher, depending on the circuit breaker frame size and trip rating. DC breakers with higher AIC ratings are available on special order.

All standard ac input breakers through 100 A trip rating are magnetic/hydraulic appliance controllers, UL recognized to standard UL 508, with an AIC of 5,000 A.

Standard ac input breakers rated over 100 A trip are thermal/magnetic breakers, UL listed to UL 489, with AIC of 10,000 A or higher, depending on the circuit breaker frame size and trip rating. AC circuit breakers with higher AIC ratings are available on special order. In addition, thermal/magnetic circuit breakers listed to UL 489 are optionally available for trip ratings under 100 A.

TRANSFORMERS   9.4.6

ANSI C57 & NEMA ST20   9.4.6.1

Transformers and inductors meet ANSI C57 and NEMA ST20 standards. All magnetic components used in HindlePower products are manufactured using a UL-recognized class 200 or class 220 insulation system, File E-75663. The design temperature rise is typically 125 °C.

Internal Wiring   9.4.7

In the SCR/SCRF battery charger, standard internal wiring is PVC to UL 1015 or equal, rated 105 °C, or XLPE[1] , rated 125 °C. Internal wiring in the AT charger series is primarily XLPE. Wire sizing is based on the NEC wire ampacity ratings but is modified where appropriate to account for the higher temperature rating of XLPE wire. Also, the NEC rates wire ampacities based on an ambient temperature of 30 °C; we adjust wire sizes to account for the higher operating ambient temperature of 50 °C.

Switchboard class wiring insulation is optionally available for the SCR/SCRF charger. This is frequently designated as *SIS* wiring. Most wire manufacturers now supply XLPE insulation to meet this requirement.

1   XLPE = Cross-linked polyethylene; PVC = Polyvinyl chloride.

APPENDIX A: FACTORY TESTING & ADJUSTMENTS

STANDARD TESTING

ALL HINDLEPOWER BATTERY chargers are fully tested before shipping, using the production test criteria of NEMA PE 5. A test report is included with each charger, along with a complete list of functional replaceable parts. A sample test report (for the SCR/SCRF charger) is reproduced in Figure A1.

The following parameters are included in each production test. Acceptance criteria are outlined in NEMA PE 5, except where in-house specifications are more stringent.

- Dielectric withstand test.
- Current limit test.
- Regulation and efficiency (voltage deviation test). During this test, the voltage adjustment range is also verified. As part of this test, the maximum ac input current is measured and recorded on the test data report. Note: the actual test value for a charger may be lower than the rated value expressed in product literature or on the charger nameplate.
- DC output ripple voltage. While this is a design test as defined by NEMA PE 5, the ripple level is verified for each charger during production testing. For convenience, the ripple may be measured at the charger output terminals, instead of the battery terminals. The ripple at the battery terminals is usually lower, due to the impedance of the wiring in the test area.
- DC voltmeter and ammeter accuracies are verified.
- All installed options are verified for proper operation and calibrated as required.
- For the AT series, the functionality of front panel controls is verified, and the HVDC Shutdown action is verified. Circuit board-level controls (such as the front panel disable jumper) are tested and verified.

Figure A1: Sample test
report for SCR/SCRF
charger

PRODUCTION TEST REPORT - SCR/SCRF BATTERY CHARGER

Sold to	Model No.		Serial No.	
DOWD BATTERY CO. INC.	1SCRF260-025-E		65819/142110-01	
Battery Type	No. of Cells		Float Voltage	Equalize Voltage
PB	116		261	270.3
Test Date	Technician		Input Voltage	
June 1, 2018	RV		240	

TEST	PARAMETER	CONDITION	DATA	UNITS
DIELECTRIC TESTS	AC input to chassis		2500	Vac
	AC input to dc output		2500	Vac
	DC output to chassis		2500	Vac
CURRENT LIMIT TEST	Output current	Vdc=203 Aac=55.9	27.5	Adc
REGULATION AND EFFICIENCY TESTS	Output voltage	264 Vac, 1.25 Adc	267.6	Vdc
	Output voltage	240 Vac, 12.5 Adc	270.3	Vdc
	Output voltage	240 Vac, 25 Adc	270.3	Vdc
	Output voltage	216 Vac, 25 Adc	272	Vdc
	Input current	240 Vac, 25 Adc	52.8	Aac
	Input power	240 Vac, 25 Adc	7550	Watts
	Voltage regulation	Combined line and load regulation	0.82	Percent
	Efficiency	240 Vac, 25 Adc	89.5	Percent
RIPPLE TEST	Ripple voltage	240 Vac, 261 Vdc, 25 Adc	57.3	mV rms

DESIGN TESTING

DESIGN TESTING IS performed on each ratings group according to NEMA PE 5.
Every rating is verified for temperature rises within design limits, and overall performance
parameters are verified for the largest rating in each enclosure, for each output voltage rating.
For example, a Style 3 enclosure accepts 130 Vdc charger ratings from 75 Adc to 125 Adc.
Each rating in that range will be verified for component temperature rises, including all
magnetics. The 125 Adc rating will be fully tested for all parameters as defined by NEMA
PE 5. Design test reports are available.

CUSTOM TESTING

TESTING TO STANDARDS OTHER THAN NEMA PE 5

Electrical testing to equipment standards other than NEMA PE 5 can be performed on special order.

BURN-IN

Burn-in of a battery charger is available. Because of limited test facilities, there may be an impact on delivery schedules. Please allow sufficient lead time, particularly if you need an extended burn-in duration. Normal burn-in periods are 24–48 hours; durations of one week or longer will certainly affect delivery.

BURN-IN WITH TEMPERATURE RISE TEST

A burn-in test can be fully instrumented to provide a report of temperature rises for critical components. These include magnetics, rectifier SCRs and/or diodes, filter capacitor terminals, and internal ambient temperature. Unless other points are specified on the order, the main transformer, (T1), main inductor (L1), and one or more SCRs will be instrumented.

The test report provides tabular data of temperature rise, and a graph of the measured components. A sample graph is shown in Figure A2.

STANDARD FACTORY SETTINGS

ALL CHARGER OPERATING parameters, such as float and equalize voltages, and most alarm and supervisory circuits can be adjusted in the field to meet the needs of the site and/or the operators. The settings available may be an actual operating voltage or a sensitivity adjustment.

AT10-130-025
Temperature Rise Test

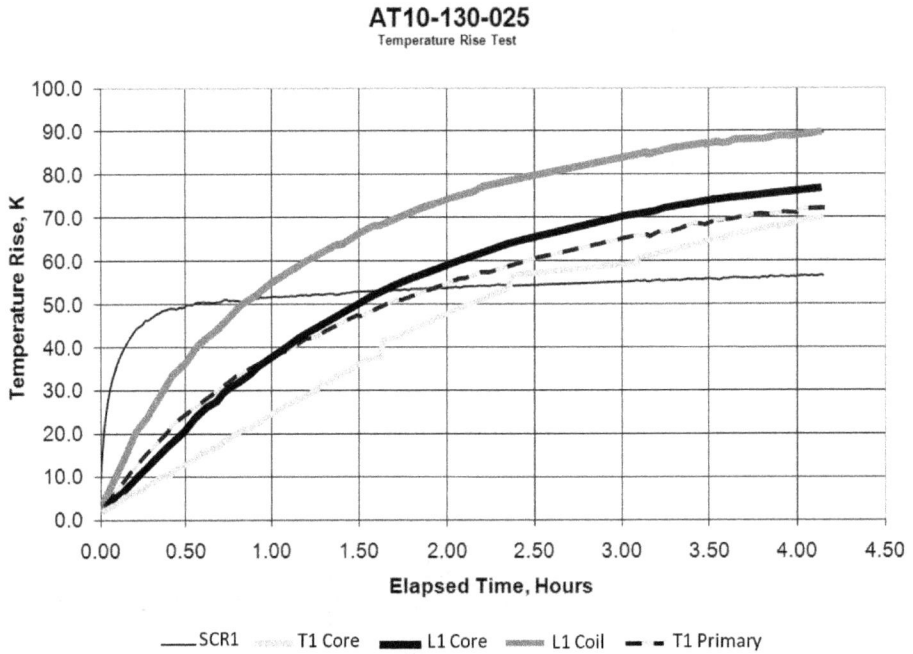

Figure A2: Sample graph of a burn-in test, showing temperature rise, in degrees Kelvin, of critical components of a battery charger

The default factory settings for the three basic parameters, set at final test, are shown in Table A1. You can override these values by specifying the settings on your purchase order or specification.

Cell Type	Float VPC	Equalize VPC
Lead-acid	2.18	2.32
NiCd	1.40	1.47

Current limit is set for all units to 110% of rating.

Table A1: Default factory settings for float and equalize voltage, based on cell type to be charged

ALARM SETTINGS — AT SERIES

The adjustment limits for the built-in alarms are given in the Specifications section of the Operating and Service Instructions. The limits are repeated here in Table A2 for convenience, along with the settings for other accessories. The upper and lower limits for the built-in alarms cannot be customized.

ALARM	12 Vdc			24 Vdc			48 Vdc			130 Vdc		
	Low Limit	High Limit	Factory Setting	Low Limit	High Limit	Factory Setting	Low Limit	High Limit	Factory Setting	Low Limit	High Limit	Factory Setting
High dc Voltage	12.0	19.0	14.4	24.0	38.0	28.8	48.0	76.0	57.6	120.0	175.0	144.0
Low dc Voltage	7.0	12.0	12.0	15.0	24.0	24.0	30.0	48.0	48.0	80.0	120.0	120.0
Charger Failure	Not adjustable by user. The charger failure alarm is activated when the charger dc output current is a predetermined value below the current limit & the dc output voltage is below the set point.											
Ground Fault	5 k	50 k	10 kΩ	5 k	50 k	10 kΩ	5 k	50 k	10 kΩ	5 k	50 k	10 kΩ
HVDC Shutdown	The HVDC Shutdown operating point is the same as the High dc Voltage alarm setting. There is a 30 second delay between alarm activation & charger shutdown when shutdown is enabled by the user.											
Low Level Detector	6.0	12.0	10.5	12.0	24.0	21.0	24.0	48.0	42.0	60.0	120.0	105.0
	The Low Level Detector (LLD) is a separate analog circuit that monitors the dc bus voltage and activates the Common Alarm relay when the dc voltage drops below the set point.											
Exhaust Fan Control	12.0	16.0	13.9	24.0	32.0	27.5	48.0	64.0	55.0	120.0	150.0	139.0
	Exhaust Fan Control should be set to operate when the battery voltage, during equalize charging, approaches the gassing threshold, about 2.3 VPC for a lead acid battery & 1.56 VPC for a NiCd battery.											

Table A2: Adjustment limits for AT Series chargers' built-in alarms and accessories

ALARM SETTINGS — SCR/SCRF SERIES

The adjustment limits for the most commonly used alarms are given in Table A3. The upper and lower limits for these alarms and accessories can be customized by special order (for example, for a custom dc output voltage).

ALARM	12 Vdc		24 Vdc		48 Vdc		130 Vdc		260 Vdc	
	Low Limit	High Limit	Low Limit	High Limit	Low Limit	High Limit	Low Limit	High Limit	Low Limit	High Limit
CASM										
High dc Voltage	11.5	17.3	23	34.6	46	69.2	115	173	230	346
Low dc Voltage	9.6	14.4	19.2	28.8	38.4	57.6	96	144	192	288
Charger Failure	In the "Zero Current" mode, can be adjusted from 10% to 2% of the rated charger output current. In the "True CFA" mode, the adjustment range is the same, but the circuit won't send an alarm unless the charger is unable to produce output current.									
Ground Fault	17 kΩ ± 20%, not user-adjustable.									
High ac Voltage	Adjustable from nominal ac input voltage to +15% of nominal.									
Low ac Voltage	Adjustable from nominal ac input voltage to -15% of nominal. In a three-phase charger, also sends an alarm for the loss of a phase.									
LEGACY ALARMS										
AC Failure Relay	Not adjustable. The alarm is typically activated when the ac input voltage drops to less than 20% of nominal, and recovers when ac voltage returns to above 70% of nominal.									
HLVA: HVDC	10.0	15.0	20.0	30.0	40.0	60.0	100	150	200	300
HLVA: LVDC	8.0	13.5	16.0	27.0	32.0	54.0	80.0	135	160	270
CFA (Charger Failure Alarm)	Despite its name, not a true Charger Failure Alarm. Responds to near zero dc current. Adjustable from 10% to 2% of the rated output current.									
Ground Detection Indicators or Relays	Not adjustable. The indicators have a sensitivity in the low thousands of ohms (2 k – 3 k). The relays have sensitivity in the low hundreds of ohms.									
High dc Voltage Shutdown	11.5	17.3	23	34.6	46	69.2	115	173	230	346
Exhaust Fan Control	Not adjustable. The exhaust fan is energized whenever the charger enters the equalize mode, whether manually or automatically.									
End-of-Discharge Alarm	9.6	14.4	19.2	28.8	38.4	57.6	96	144	192	288
Battery Discharge Alarm	This option is normally rated for the maximum dc load current seen by the battery. Adjustable from 10% to 2% of the rated load.									

Table A3: Adjustment limits for SCR/SCRF Series chargers' built-in alarms and accessories. Note: This table doesn't show a factory setting because the value depends on the number of battery cells specified on the order

APPENDIX B: USEFUL DATA

TEMPERATURE & ALTITUDE DERATING

IN THE SECTION on *Packaging* (Section 2.5), we mentioned, almost in passing, that the rating of a charger has to be adjusted (increased) if it will be utilized outside of certain temperature limits, or above a certain elevation.

Chargers are designed to operate continuously, at full output power, at a maximum ambient temperature of 50 °C (122 °F), but this rating applies at sea level to an elevation of 1,000 m (about 3,000 ft). If the charger is to be operated outside those limits, you have to adjust its output rating.

All electronic components generate some heat during operation. The electronic components inside the charger depend on air convection cooling to keep their temperatures within safe limits.

SCRs (semiconductor controlled rectifiers) can operate safely, at their maximum ratings, with a stud temperature of around 100 °C. The SCR is mounted on an aluminum heat sink that must be designed to allow no more than a 50 °C rise in temperature with the SCR operating at full tilt. [Actually, the design target is a 40 °C rise, giving a 10 °C margin for stud temperature.]

You can see that if a charger is operated at an ambient temperature over 50 °C, the SCR could be overstressed. We can compensate for this by reducing the maximum output current of the charger so that the SCR current is proportionally lower, reducing the temperature rise. This is what we mean by derating.

The problem of overheating can be compounded by the fact that air density decreases a little at the higher temperature. Air density also decreases with elevation, but this is slightly offset by the fact that air temperature also decreases with elevation. At 3000 meters (about 10,000 feet), the air temperature will normally be about 20 °C cooler than at sea level, but the pressure and density will be 30% lower. So the cooling effectiveness of air is lower at higher elevations.

At high temperatures, or high elevations, you must order a charger with a higher rating than would normally be required for the installation. You can't work in an environment higher than 85 °C, however; in fact, at that temperature, a charger rating is zero. You also shouldn't work at an elevation higher than 3,000 meters.

To choose a charger rating to work at a temperature between 50 °C and 85 °C, use the following formula[1]:

$$Adc = -I_0 * \left(\frac{35}{85 - T}\right)$$

where Adc is the new charger rating, I_0 is the desired output current, and T is the maximum operating temperature.

Note that this expression is valid only between 50 °C and 85 °C. You should round the required current up to the next available charger rating (e.g., if the calculation results in a charger rating of 90A, round up to 100A, the next available rating).

The standard charger rating is good to an elevation of 1,000 m (about 3,300 feet). If you need to rate a charger for higher elevations, use the following formula:

$$Adc = -I_0 * \left(\frac{2000}{3000 - H}\right)$$

where, as above, Adc is the new charger rating, and I_0 is the desired output current. H is the elevation in meters; if you know the elevation in feet, divide that number by 3.28 to get meters.

Things get a little more complicated if you have a combination of high elevations and high temperatures. Check with your manufacturer if this situation applies to you. The HindlePower web site has a helpful application note in the Support | Technical FAQs section, named, appropriately, Altitude and Temperature Derating. This note provides a graph that combines the derating factors for combinations of higher temperatures and higher elevations.

---

1   Both formulas for derating on this page were derived by the author.

RECOMMENDED SPARE PARTS

IN A PERFECT world, nothing breaks or fails prematurely or wears out. For this reason, manufacturers usually recommend that users stock a small selection of components as spare parts.

The instruction manual shipped with every HindlePower charger contains one or more tables listing all of the replaceable electrical and electronic parts. A subset of these parts are noted as "recommended spares." For the AT series, the parts recommended as spares are noted in the master table of part numbers. For the SCR/SCRF series, there is a separate table (Section VI), describing each component and noting whether it's a recommended spare.

Note that for the SCR/SCRF series, we describe two classes of spares: start-up and operating. There is little difference; those parts labeled as start-up spares are the ones most likely to have a problem in the event of any malfunction during installation, start-up, or commissioning of a charger, such as fuses.

Operating spares can cover a lot of ground, and include control circuit board(s), SCRs, (or complete rectifier assemblies), filter capacitors, and the like. These are the parts that are most likely to contribute to an early failure (known as "infant mortality"), or fail as the result of an exceptional operating condition, such as a bus fault, or severe lightning strike. These are the parts to have around if your application is critical, and downtime must be minimized.

How likely are you to have an early failure? Not very much. And a truism about electronic equipment is that if it operates satisfactorily for the first few weeks, you're probably golden for many years.

TECH TIP

STANDARD PARTS LISTS

Each charger is shipped with an instruction manual (officially, the Operating and Service Instructions), and also a separate sheet called a Parts Data Package (PDP). The PDP lists all replaceable parts for that specific charger, and notes which ones are recommended as spare parts. If the purchase order specifies a "priced parts list," the PDP also shows the list price for each component; the price is guaranteed for one year from the date of the shipment.

CUSTOM PARTS

Some charger orders require substituting standard parts with custom components. An example would be a custom dc output voltage, available with the SCR/SCRF charger series. Components such as R1 (bleed resistor), control loop resistors, and scaling resistors on alarm boards, such as the CASM, need to be adjusted. In some cases, filter capacitors need to be changed. For a case like this, the Parts Data Package is automatically revised to include the custom parts. Users should always consult the PDP when a replacement part is required.

TECH TIP

PARTS DATA PACKAGE

Figure B1 shows the heading and first few rows of a Parts Data Package. The serial number uniquely identifies each charger built. The rows for each component show the Reference Designator (aka "Circuit Symbol"), the factory part number, and a short description. Whether the part is recommended as a spare is shown in the last three columns.

PARTS DATA PACKAGE

Sold to: Ship to:

Model:	AT30130200F480SXHXAXXXXX		Ser. No.:			
Factory Configuration No.:	A3130200F480SXHXAXXXXX					
	EJ5056-52	ASSY,MAINFRAME,AT30,130V,200A				
	EJ5057-52	OPT,DC.FILTER,STD,AT30-130-200				
	EJ5060-52	OPT,480VAC,AT30-130-200				
	EJ5066-52	OPT,CB1-8TD,480V,AT30-130-200				
	EJ5082-52	OPT,CB2-HAIC,AT30-130-200				
	EJ1156-03	OPT,AUX.RELAY,PCB,AT30,STYLE-5030				
	EJ1054-52	OPT,LABEL,AT30,				

Reference Designator	Factory Part Number	Description	Quantity per Unit	Start-up Spares	Operating Spares	PM Spares
A1	EN5002-00	ASSY,PC.BOARD,AT.SERIES,MAIN.CTRL	1	Yes	Yes	
A15	EN5008-04	ASSY,PCBD,GATE.DRIVER,AT30,130V	1			
A16	PM5007-07	ASSY,MOD,SCR/HS,3PH,800V,275A	1			
A5	EN0027-00	ASSY,PC BRD,AT SERIES,AUX.RELAY,A5	1			
C1	RP0019-08	CAP,ELEC,2.5"CAN,200V,7400 UF	4			Yes
CB1	RE0016-12	CKT.BKR,CD,3P,480V,70A	1			
CB2	RE5038-04	CKT.BKR,SQ-D,PP,JD,2P,600V,250A	1			
CR1	RK0017-14	DIODE,RECT,3/4",300A,400V,REV POL	1	Yes		

Figure B1: Example Parts Data Package for an AT charger

An actual PDP contains many more rows, covering every replaceable component in the charger. The document finishes at the bottom with a couple of messages concerning pricing, and a note advising the user of any possible custom parts.

» *Q: Why can't I just get the Parts Data Package from your web site?*

A: Web site documents are standard drawings and procedures. Each PDP is unique to a specific charger. Any web site document wouldn't apply specifically to your charger.

BATTERY SIZING FOR A COMPLEX LOAD PROFILE — IEEE 485 EXAMPLE

UNDER THE TOPIC, *Discharging Stationary Batteries* (Section 1.4), we discussed sizing a lead-acid battery for a complex load profile (Figure B2).

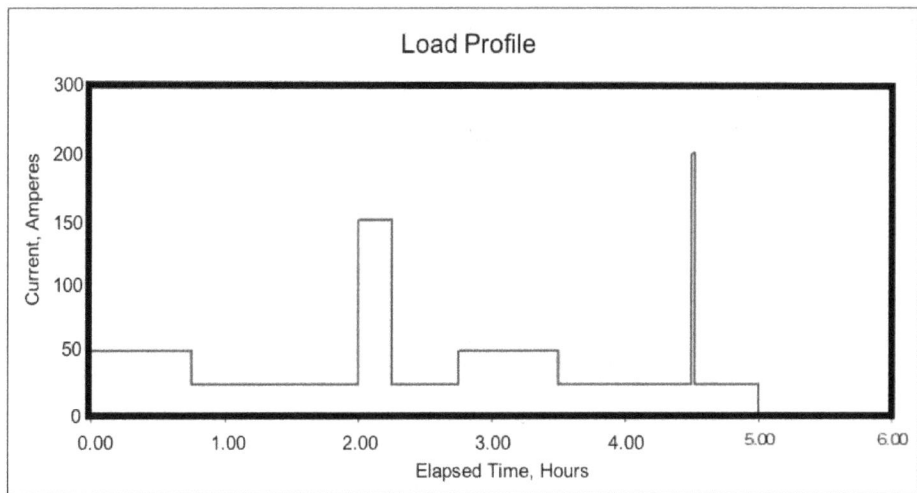

Figure B2: Complex load profile used in the calculations to size a lead-acid battery using a method based on the original Hoxie Method

The following is an example of sizing the battery using manual calculations with the eight periods of load current in this profile according to IEEE 485. The IEEE standard outlines two calculation approaches to sizing the cell. We'll illustrate the method using the R_t cell capacity factor. (Note that IEEE-485 contains a wealth of other application information, including a rigorous mathematical basis for cell sizing.)

Table B1 is a simplified version of the worksheet provided by IEEE-485. The calculations for each load period can be found after this table.

Period	Load, Adc	Change in Load, A	Duration of Period, min.	Time to End of Section	Capacity A/Pos.	Number of Positive Plates*	
						Pos. Values	Neg. Values
Section 1: First Period (If A2 is greater than A1, go to Section 2)							
1	A1=50	A1-0=50	M1=45	T=M1=45	45	1.11	
					Section 1 Total		
Section 2: First two periods only (If A3 is greater than A2, go to Section 3)							
1	A1=	A1-0=	M1=	T=M1+M2=			
2	A2=	A2-A1=	M2=	T=M2=			
					Section 2 Subotal		
					Section 2 Total		
Section 3: First three periods onlyl (If A4 is greater than A3, go to Section 4)							
1	A1=	A1-0=	M1=	T=M1+M2+M3=			
2	A2=	A2-A1=	M2=	T=M2+M3=			
3	A3=	A3-A2=	M3=	T=M3=			
					Section 3 Subtotal		
					Section 3 Total		
Section 4: First four periods only (If A5 is greater than A4, go to Section 5)							
1	A1=	A1-0=	M1=	T=M1+M2+M3+M4=			
2	A2=	A2-A1=	M2=	T=M2+M3+M4=			
3	A3=	A3-A2=	M3=	T=M3+M4=			
4	A4=	A4-A3=	M4=	T=M4=			
					Section 4 Subtotal		
					Section 4 Total		
Section 5: First five periods only (if A6 is greater than A5, go to Section 6)							
1	A1=	A1-0=	M1=	T=M1+M2+M3+M4+M5=			
2	A2=	A2-A1=	M2=	T=M2+M3+M4+M5=			
3	A3=	A3-A2=	M3=	T=M3+M4+M5=			
4	A4=	A4-A3=	M4=	T=M4+M5=			
5	A5=	A5-A4=	M5	T=M5=			
					Section 5 Subtotal		
					Section 5 Total		
Section 6: First six periods only (if A7 is greater than A6, go to Section 7)							
1	A1=	A1-0=	M1=	T=M1+M2+M3+M4+M5+M6=			
2	A2=	A2-A1=	M2=	T=M2+M3+M4+M5+M6=			
3	A3=	A3-A2=	M3=	T=M3+M4+M5+M6=			
4	A4=	A4-A3=	M4=	T=M4+M5+M6=			
5	A5=	A5-A4=	M5=	T=M5+M6=			
6	A6=	A6-A5=	M6=	T=M6=			
					Section 6 Subtotal		
					Section 6 Total		
Section 7: First seven periods only (if A8 is greater than A7, go to Section 8)							
1	A1=	A1-0=	M1=	T=M1+M2+M3+M4+M5+M6+M7=			
2	A2=	A2-A1=	M2=	T=M2+M3+M4+M5+M6+M7=			
3	A3=	A3-A2=	M3=	T=M3+M4+M5+M6+M7=			
4	A4=	A4-A3=	M4=	T=M4+M5+M6+M7=			
5	A5=	A5-A4=	M5=	T=M5+M6+M7=			
6	A6=	A6-A5=	M6=	T=M6+M7=			
7	A7=	A7-A6=	M7=	T=M7=			
					Section 7 Subtotal		
					Section 7 Total		
Section 8: First eight periods only (if A9 is greater than A8, go to Section 9)							
1	A1=	A1-0=	M1=	T=M1+M2+M3+M4+M5+M6+M7+M8=			
2	A2=	A2-A1=	M2=	T=M2+M3+M4+M5+M6+M7+M8=			
3	A3=	A3-A2=	M3=	T=M3+M4+M5+M6+M7+M8=			
4	A4=	A4-A3=	M4=	T=M4+M5+M6+M7+M8=			
5	A5=	A5-A4=	M5=	T=M5+M6+M7+M8=			
6	A6=	A6-A5=	M6=	T=M6+M7+M8=			
7	A7=	A7-A6=	M7=	T=M7+M8=			
8	A8=	A8-A7=	M8=	T=M8=			
					Section 8 Subtotal		
					Section 8 Total		

* Calculated as the change in load ÷ plate capacity at T minutes

Table B1: Calculations for the eight periods of the load profile shown in Figure B1

First, we construct Table B2 which summarizes the current and duration for the eight periods of load current for the profile shown in Figure B1.

Table B2: Current and duration for the eight periods of load current that make up the load profile shown in Figure B1

Period	Total Amperes	Duration, Minutes
1	50	45
2	25	75
3	150	15
4	25	30
5	50	45
6	25	60
7	200	01
8	25	29

In Table B3, we consider the load currents in the first two periods. The period 1 load current is greater than that of period 2, so we enter the data in the first section of the worksheet.

Period	Load, Adc	Change in Load, A	Duration of Period, min.	Time to End of Section	Capacity A/Pos.	Number of Positive Plates*	
						Pos. Values	Neg. Values
Section 1: First Period (if A2 is greater than A1, go to Section 2)							
1	A1=50	A1-0=50	M1=45	T=M1=45		1.11	
					45		
					Section 1 Total	1.11	

Table B3: Worksheet calculations for the first period of the load profile shown in Figure B1

We'll use the capacity data graph for the "QZN" cell shown in IEEE-485. We calculate that we need 1.11 positive plates to meet the capacity requirement for the first period. But we aren't done.

Period 2 has lower load current than period 1, so we skip to section 3 of the worksheet. We enter the data for the first three periods as seen in Table B4 (section 2 is left blank). Note that the incremental number of positive plates for period 2 is negative, since the load current is lower than for period 1.

Section 3: First three periods only! (if A4 is greater than A3, go to Section 4)						
1	A1=50	A1-0=50	M1=45	T=M1+M2+M3=135	25	2	
2	A2=25	A2-A1=25	M2=75	T=M2+M3=90	32		0.78
3	A3=150	A3-A2=125	M3=15	T=M3=15	70	2.14	
					Section 3 Subtotal	4.14	0.78
					Section 3 Total	3.36	

Table B4: Worksheet calculations for the first three periods of the load profile

The current for period 4 is lower than for period 3, so we skip to section 5, as we did for section 2. Enter the data for the first 5 periods in section 5 as seen in Table B5.

Section 5: First five periods only (if A6 is greater than A5, go to Section 6)						
1	A1=50	A1-0=50	M1=45	T=M1+M2+M3+M4+M5=210	18	2.77	
2	A2=25	A2-A1=25	M2=75	T=M2+M3+M4+M5=165	24		1.04
3	A3=150	A3-A2=125	M3=15	T=M3+M4+M5=90	32	3.9	
4	A4=25	A4-A3=-125	M4=30	T=M4+M5=75	36		3.47
5	A5=50	A5-A4=25	M5=45	T=M5=45	45	0.55	
					Section 5 Subtotal	7.22	4.51
					Section 5 Total	2.71	

Table B5: Worksheet calculations for the first five periods of the load profile

Having fun? You may be wondering where all these numbers come from. The first 4 columns are self-explanatory. Column 5, the time to the end of the section, is a summation of all the section times for each period and expresses the duration of each current increment in the section. For example, look at period 3 above. The current increment is 125A (A3 – A2), and the total time that current flows is for periods 3, 4 and 5, or 90 minutes. For column 6, we consult the sample battery curve (from IEEE-485) to determine the amperes we can get from a single plate for that total time, 90 minutes. The number of plates we need for any period is column 3 divided by column 5 (negative if the load change is negative). The reason for all this, remember, is that higher load currents reduce battery capacity faster than lower load currents; this method accounts for those differences.

Period 6 has a lower current than period 5, so we would normally jump to section 7. Note, however, that period 7 is a large momentary load – we rate it for 1 minute, even if it's only a few seconds long (as for tripping a breaker, for example). We then see that period 8 is much smaller, at 25A, than period 7. This would normally require us to jump to period 9, which doesn't exist. How do we handle this? IEEE-485 doesn't provide any guidance.

The answer is to move the momentary load to the end and treat the smaller 25A load as if it were period 7. This is only a little bit conservative; it might add a few Ah to the battery size.

Now we have the revised summary Table B6 (since period 6 and period 7 are now the same current, I've folded them together as the new period 6):

Table B6: To simplify calculations, period 8 from Table B2 is moved up one row and combined with period 6

Period	Total Amperes	Duration, Minutes
1	50	45
2	25	75
3	150	15
4	25	30
5	50	45
6	25	89
7	200	01

So we skip now to section 7 of the worksheet, as shown in Table B7.

Section 7: First seven periods only (if A8 is greater than A7, go to Section 8)							
1	A1=50	A1-0=50	M1=45	T=M1+M2+M3+M4+M5+M6+M7=300	14	3.57	
2	A2=25	A2-A1=25	M2=75	T=M2+M3+M4+M5+M6+M7=255	17		1.47
3	A3=150	A3-A2=125	M3=15	T=M3+M4+M5+M6+M7=180	22	5.68	
4	A4=25	A4-A3=-125	M4=30	T=M4+M5+M6+M7=165	24		5.21
5	A5=50	A5-A4=25	M5=45	T=M5+M6+M7=135	26	0.96	
6	A6=25	A6-A5=-25	M6=89	T=M6+M7=90	32		0.78
7	A7=200	A7-A6=175	M7=1	T=M7=1	101	1.73	
					Section 7 Subtotal	11.94	7.46
					Section 7 Total	4.48	

Table B7: Worksheet calculations for the seven periods of the load profile (modified from the original eight periods in Figure B2)

The final total for this hypothetical cell is 4.48 plates (round up to 5 positive plates). Since, from the sample curve, the plate is rated 80 Ah at the eight-hour rate, this load profile needs a 400 Ah battery.

You will note in IEEE-485 that there are other approaches you can take to the calculation, such as determining the ampere hour requirement for the battery instead of the number of positive plates. The various methods should yield the same result.

Or not. The calculations above don't take into consideration the operating temperature range, battery aging, or design margins. You should repeat the calculations for the worst-case temperature conditions, or adjust the final result for these variables. Also, you might repeat the calculation using one or two other cell sizes, to see if you can achieve a more economical solution. But your best bet is to contact the manufacturer and use their battery sizing software.

CALCULATING BATTERY INTERNAL RESISTANCE

IN THE DISCUSSION of short-circuit current (Section 1.4.8), we gave a rule of thumb for estimating the short-circuit current that could be obtained from a lead-acid battery, namely, that the initial current could be about 10.2 times the 1-minute discharge rate, based on an end voltage of 1.75 VPC. This gives the value for 25 °C; the resulting current can be 6 to 10% higher if the electrolyte is at a higher temperature. The multiplier assumes an initial specific gravity of 1.210, and that the battery is at equalize charge, 2.33 VPC; use a factor of 9.8 if the battery is on float, at 2.25 VPC.

Alternatively, you can use a factor of 10 times the Ah rating of the battery (at the eight-hour rate). A 200 Ah battery, for example, could deliver 2,000 A or more into a high-quality short circuit. This calculation usually results in a lower estimated current than the first estimation.

While these rules of thumb may get you close, it may not be close enough for determining the system protection required for a fault on the dc bus. Normally, dc fuses are included in the battery connections as part of the overall system fault protection. IEEE Standard 1375 has an extensive discussion of dc fuse selection and application. The standard also has an example of calculating the internal resistance of a battery, using the battery manufacturer's discharge characteristic curves. You use the internal resistance to calculate the prospective initial short-circuit current into a load fault.

Fault protection can be complex, and I urge you to get familiar with IEEE 1375.

If you need to know the internal resistance of a battery, the easiest way to find it is to ask the manufacturer. Failing this, you can calculate the internal resistance from performance curves provided by most manufacturers. You need to use the curve that plots initial discharge voltage vs. discharge current. Although manufacturers present the data in various ways, the curve might look like the one shown in Figure B3.

Note that the curve relates the terminal voltage of a cell to the load current, in amperes per positive plate. In most constructions, there are an odd number of plates, since there is one more negative plate than positive plates. A hypothetical battery with a designation such as QZN-7, for example, would have 3 positive plates and 4 negative plates. The total

TECH TIP

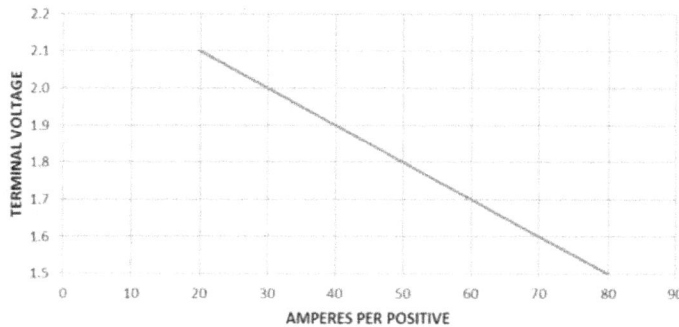

Figure B3: Example discharge voltage versus discharge current plot that may be provided by a battery manufacturer which can be used to calculate its internal resistance

battery current capability, of course, is the amperes per positive plate multiplied by the number of positive plates.

You calculate the internal resistance of a one positive plate cell by determining the slope of the discharge curve. In this example, find two points on the curve that give a difference of a few tens of amperes. We'll use the points (2.0 V, 30 A) and (1.8 V, 50 A). Now use the formula

$$Ri = \frac{V_1 - V_2}{I_2 - I_1}$$

where R_i is the internal resistance of the cell. With the numbers above, the resistance will be 0.01 ohm (which is a high cell resistance). For the QZN-7 in this example (3 positive plates), the total resistance is 0.01 ÷ 3, or about 3.3 mohms. This is the internal resistance for one cell. The short-circuit current available from that cell, assuming an initial open-circuit voltage of 2 V, is about 606 amperes. If you have several cells in series, say 60 cells for a 120 V battery, the current would be the same, since the cell resistances add algebraically.

APPENDIX C: SOLVING PROBLEMS

MAINTENANCE — PREVENTIVE & OTHERWISE

YOU PROBABLY HAVE a regular maintenance schedule for your battery installations. Do you need to concern yourself with maintenance on the charger(s) and other dc power equipment in the system?

The short answer is yes. While you don't have the same responsibilities as for electromechanical equipment such as generators or M-G sets, it's important to pay attention to housekeeping and certain critical subsystems.

Ensuring that a charger is cooled properly is important. Spiders love to build their webs across cooling vents, and even within the fins of heat sinks. We recommend that you clean these passageways at least annually.

If your charger is fan-cooled, this is a convenient time to clean the fans and check for proper operation. Enclosure-mounted fans (used on some large chargers, and in sealed enclosures like NEMA 4 or NEMA 12) are equipped with filters. Be sure the filters are unclogged. The fans used in HindlePower chargers use ball bearings exclusively and should have a service life of many years. In case of any trouble, fan-cooled chargers are equipped with over-temperature sensors to warn of fan failure.

The Operating and Service Instructions manual that is shipped with the charger contains a sample preventive maintenance procedure, and a form for recording the maintenance history on your equipment.

TECH TIP

PERIODIC INSPECTIONS & SCHEDULED MAINTENANCE

As noted above, clean all cooling vents and fan filters at least semi-annually. If equipment is in an abnormally dusty environment, this needs to be done more frequently.

Check the wiring connections at the input and output terminals to be sure they are tight. This should be done at least once a year. If you have any alarm or annunciator connections, check them for tightness, also.

Test the front panel indicators. The AT series has a Lamp Test button on the front panel. Some options in the SCR/SCRF series (such as the CASM) also have lamp test functions. If they don't, one way to test the lamps is to induce an alarm condition (for example, to check a Charger Failure alarm, you can simply turn off the charger (*be sure the battery is connected in the system!*) for a few seconds.

Finally, if a charger is filtered, measure the ripple voltage, preferably at the battery terminals, using a digital voltmeter set to measure millivolts ac. A very high reading (well over 100 mv) may indicate a problem with the dc filter capacitors.

ELECTIVELY REPLACING ELECTROLYTIC FILTER CAPACITORS

Now, about those filter capacitors. They're the large cylindrical cans mounted on the main panel inside the charger. Historically, we have recommended that they be electively replaced after 10 years of service, since the internal resistance of electrolytic capacitors increases, and the capacitance decreases, gradually over time.

Electrolytic capacitors in HindlePower products are generously rated. In general, the working (dc) voltage is limited to 75% of the capacitor rating, and the ripple current, on average, is no more than 50% of the specified maximum. But a major factor in capacitor life is the operating temperature. With the design margins mentioned here, internal temperature rise is about 15 °C (based on actual factory qualification testing). Most filter capacitors are rated for 85 °C operation, so that at 50 °C ambient, internal temperature can rise to 65 °C and still allow a 20 °C margin over the capacitor rating.

The effect of ripple current is more complicated, since it involves the gradual increase in ESR (effective series resistance, or internal resistance). As the ESR increases, internal temperature also rises. With the operating conditions noted above, capacitor life could be as long as 20 years (in a temperature-controlled environment), but the ESR increase is a limiting factor. And, of course, it's a regenerative process: Higher ESR means higher temperature rise, etc.

TECH TIP

There isn't any magic about that 10-year mark. As capacitors age, the ripple voltage in the dc output of the charger gradually increases. Your best course, to achieve a balance between maintenance costs and performance, is to measure the ripple voltage (annually would be fine), preferably at the battery terminals, and replace the capacitors when the ripple exceeds your comfort level. Remember that when one capacitor starts to fail, the ripple current in the remaining capacitors will increase, leading to a "domino effect," and ultimate failure will be accelerated.

» *Q: You say I should replace the filter capacitors after 10 years. Why don't I just order an unfiltered charger and save myself the trouble and expense? The battery does lots of filtering.*

A: In the past, utilities bought unfiltered chargers because the connected loads could tolerate the normal ripple voltage appearing at the battery terminals. Legacy equipment could even handle the high ripple applied by the charger when the battery was disconnected for maintenance. Traditionally, batteries were of the vented type, with enough excess electrolyte to handle the ripple current between maintenance inspections.

Substations have evolved, however. Switchgear is now electronically controlled, and the electronics can be damaged by the high ripple values that can appear if a battery is disconnected, or loses its filtering ability due to aging. Add to that the trends toward using VRLA batteries for substations, and reduced maintenance schedules, and it just makes more sense to install a filtered charger. Avoid the headaches.

GETTING MORE INFORMATION

WHILE WE'VE TRIED to anticipate all your questions in this book, we know that you may want to look deeper into some topics or have specific questions that require more research. We list here some useful sites on the internet that may help.

The official HindlePower Inc. website (www.hindlepowerinc.com) has everything you always wanted to know about HindlePower Products, including a vast treasure of application notes and service procedures: [1]

---

[1] Document numbers for various notes and procedures are sprinkled throughout this manual. They may occasionally change when a document is updated. Consult the web site (or use the Chat provision) to be sure you have the latest information.

If you need quick access to catalog data and/or drawings for the AT series, then try the dedicated AT Series website (www.atseries.net).

Both www.batteryuniversity.com and www.wikipedia.org offer educational material on battery construction, chemistry, and applications.

If you aren't familiar with Battcon™, the International Stationary Battery Conference, check out their web site at www.battcon.com. Their conference papers for the last several years are available in PDF format.

Finally, there is an abundance of useful manufacturers' information; we cannot possibly list them all, and I apologize if I omitted your favorite. Here are a representative handful of useful and informative sites:

For good application notes on electrolytic capacitors, try www.nichicon-us.com.

Comprehensive data for primary cells and small rechargeables are at www.duracell.com and www.energizer.com.

Information on large secondary batteries, both lead-acid and NiCd, can be found at www.sbsbattery.com, www.enersys.com, www.cdtechno.com, and www.alcad.com.

And don't forget the industry standards which were discussed in CHAPTER 9. In particular, check out www.ul.com, standards.ieee.org, and www.nema.org/standards.

CONTACTING CUSTOMER SERVICE

WE TEND TO think of Customer Service as the people we reluctantly call when the product doesn't work, it's damaged in shipment, or parts are missing. Or, worse, it's the wrong product or was shipped to the wrong site.

Contacting the Customer Service team is your first step in getting things made right. You may be surprised at how fast they can fix some of the problems just mentioned.

Customer Service can be a lot more than just a problem-solver, though. They're available to answer application questions, review customer specifications, provide drawings and other documentation, arrange for spare parts delivery, and, not least, work with you to customize a charger, or provide a completely customized dc power system.

You can get a rapid response from Customer Service by using the Chat function on the HindlePower web site (during normal business hours, Eastern Standard Time). Just type in your question or concern, and you'll get a response within a minute or two.

Some of the other services that Customer Service normally provides:

- Receive a lead-time report for standard delivery schedules.
- Receive a shipping report when your product leaves the factory.
- Arrange for start-up services, on-site training, or other assistance.
- Issue an RMA (Returned Material Authorization) to return a product for any reason, including the need for factory service.
- Arrange for field service (a very rare occurrence, of course).

There are a few things you should have ready before placing a call to Customer Service:

- Complete serial number of the battery charger, or other equipment.
- Site name.
- Date of installation and/or commissioning (if the equipment is in operation).
- If the equipment is powered by the ac mains, the actual ac input voltage may be helpful. If it's a charger, the Ah rating of the battery is helpful.
- Telephone number that can be reached by Customer Service during normal working hours.
- Description of the problem or other issue that needs attention.

FIELD MODIFICATIONS

CHANGING THE AC INPUT VOLTAGE

» *Q: I ordered the wrong ac input voltage. Do I have to return the charger to the factory?*

    A: In most cases for the AT series, you don't have to return the charger. The AT10.1 Group 1 single-phase charger, when ordered for 120, 208, or 240 Vac input, can be converted in the field to any one of those ac input voltages, and retains all compliance with industry standards. Note that this covers the ratings 6 Adc to 25 Adc.

    The AT10.1 Group 2 chargers (30 Adc to 100 Adc) aren't quite as flexible. The ac input can be changed from 240 Vac to 208 Vac, or vice-versa. But if you ordered 240 Vac and need

120 Vac (or the reverse), then the main transformer needs to be changed. We recommend a return trip to the factory.

The same is true for any change to or from 480 Vac input. This also requires a transformer change, and also changing to the properly-rated input circuit breaker.

Things get a little more complicated with the SCR/SCRF series. The change from 208 Vac to 240 Vac, or the reverse, is possible. For some single-phase ratings, changing from 120 Vac to 240 Vac, or the reverse, is possible in the field, but requires changing the ac input circuit breaker. We recommend returning the charger to the factory for such a change, and also for any change involving 480 Vac input.

GROUPING SELECTED ALARM RELAYS — THE THERIAULT CONNECTION

» **Q: I have an SCRF charger with a CASM alarm board installed. However, I only want to use the Charger Failure and the AC Failure alarms in my common alarm connection. How do I defeat the other alarms?**

A: It isn't practical to disable individual alarms on the CASM. It's easy, however, to create your own common alarm connection by carefully "daisy-chaining" the desired alarm contacts, according to a method invented by Matt Theriault.

In Figure C1, we show 4 alarm relays chained together, but you can do this with any 2 or more relays. Note that the output of the combined connection has Common, Normally Closed (NC) and Normally Open (NO) terminals, the same as for any individual relay.

The common output connection is wired first to the Common contact of the first alarm you want, then the NC terminal of that relay is wired to the Common contact of the next relay, and so on. The NC terminal of the last alarm relay becomes the NC output of the new common alarm. The NO terminals of the alarm relays are bused together to form the NO output of the new common alarm. Important: The relay contacts are shown in the non-alarm condition, which is how they're labeled on the CASM circuit board.

FAQS

Figure C1: Four alarm relays "daisy-chained" together. Circuit designed by Matt Theriault of HindlePower Inc.

Notes:
1. Contacts shown in non-alarm condition.
2. Contacts wired as shown are dedicated to the common alarm function, so they may not be used as individual alarm contacts.
3. Additional relays may be added in the same manner without limit.

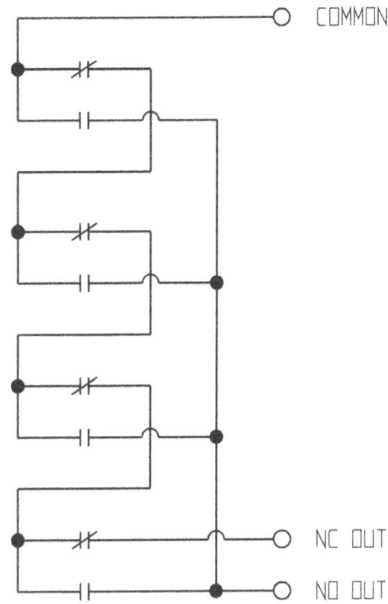

» **1) I connected my new AT10 charger to 120 Vac, following your manual exactly. There's no output current at all, although the front panel meter reads the battery voltage. What's wrong?**

The AT10.1 Group 1 ratings (6 Adc through 25 Adc) are automatically wired at the factory for 240 Vac input. Unless you specified 120 Vac on your purchase order, the charger was shipped wired for 240 Vac. Section 1.6 of the Operating and Service Instructions gives detailed instructions for changing the transformer taps for 120 Vac operation.

» **2) I need a blocking diode. Does that mean that I have to buy the SCRF charger?**

The SCR/SCRF charger series includes a blocking diode in all ratings. The original design included a blocking diode to satisfy customer specifications, which frequently specified a means to "prevent the charger from discharging the battery during an emergency." Some charger designs include "loading resistors" to ensure proper operation at light load; these are unnecessary in an SCR phase-controlled charger.

The AT10.1 Group 2 and the AT30 series don't need a blocking diode to avoid discharging the battery during an emergency. Note, however, that all chargers take some power from the battery during an emergency, to power the control and alarm circuits. Also note that the AT10.1 filtered chargers do have a blocking diode.

Unless your customer is adamant about a blocking diode, rest easy with the AT series.

» **3) Can I power my battery charger from a UPS? How about from a constant voltage transformer?**

We cover operation on a UPS in Section 2.4.6 *in response to the question, "What Alternate Sources of AC Power Can I Use with a Battery Charger?" Some of the same comments apply to operation on a constant-voltage transformer (sometimes called a line conditioner). A properly coordinated system should work OK, but CV transformers are limited in their output current capability and may have trouble supplying the inrush current needed to start a charger from scratch. This means the charger may take longer to start up.*

» **4) Why do you specify a specific startup sequence using the circuit breakers?**

Back in the dark ages of charger design (i.e., before the AT series was available), the SCR/SCRF product had a standard configuration consisting of a two-pole dc fuse for output protection. The dc circuit breaker was an option. It still is, but the vast majority of customers order the breaker, so it looks like a de facto standard. Even when the dc breaker is ordered, there is still one output fuse in the positive leg. Those output fuses are the fast-acting "rectifier" type.

When a user installs a dc system, the battery is usually not fully charged; it's close to its nominal 2.0 VPC (for lead-acid batteries). The charger, if in float mode, is ready to deliver about 2.25 VPC to 2.30 VPC. If you turn on the ac breaker first, you charge the dc filter capacitors (in a filtered charger), but then the capacitors will be at 2.3 VPC when you turn on the dc breaker. Since soft start is over, the charger is ready to deliver whatever current is necessary to bring the battery up to float voltage. Of course, there's the current limit circuit, but it won't limit the discharge current from the filter capacitors, which may deliver just enough current to clear that pesky fuse.

Does it happen often? No, and there are no guarantees. Probably there's a little more risk with a battery eliminator filter. We developed the startup sequence because it offered a more reliable way to get the system going. Why doesn't the fuse clear when you turn on the dc breaker first? Because a filtered SCR/SCRF charger has a blocking diode, and the battery can't provide any capacitor charging current.

What about the AT product? The AT design uses a slower fuse, which is possible because the AT charger has a faster-acting current limit circuit, and also has the current crowbar circuit to protect against events like dc bus faults.

GLOSSARY

This section contains definitions of terms and concepts used throughout the book. Terms that are used primarily in a specific section (such as ripple) are defined in that section.

Ampere hour	Most common measure of the capacity of a cell or battery. A cell that can deliver one ampere for one hour, while maintaining its terminal voltage above a certain value (which depends on the cell type), has a capacity of one ampere hour.
Anode	The terminal of an electronic device *from which* electrons enter the external circuit. This term causes a lot of confusion because the anode is the positive terminal of a device like a diode, but the negative terminal of a primary battery. If you can handle electrons, the definition above is the most accurate.
Battery	Traditionally, an assembly of two or more electrochemical cells for storing electrical charge and delivering that charge in the form of electrical current to an external circuit. Because *battery* is widely used, informally, to refer to single *cells* (such as flashlight or camera *batteries*), we'll do the same. We may use the terms *battery* and *cell* interchangeably.
Cathode	The terminal of an electronic device that collects electrons from the external circuit. The cathode is the positive terminal of a primary cell.
Capacity	Measure of a battery's stored charge and the useful electrical current that it can provide over time to an external circuit. Since it is the product of the current drawn from the battery, multiplied by the hours the current flows, it is expressed in ampere hours, or Ah.
Cell	The basic unit of an electrochemical battery, consisting of one anode, one cathode, electrolyte, and whatever additional materials or parts are required to hold it all together, foster the electrochemical reaction(s), and either deliver electrical current to an external circuit, or return current to the cell to replenish the stored charge.
Cycle Life	A rechargeable battery, after a complete discharge and subsequent recharge, has completed one cycle. Cycle life is defined as the number of complete cycles that a battery can endure before the discharge capacity decreases to 80% of its rating.

Diode	A device that conducts electrical current in only one direction, like a check valve. Handy for changing alternating current into direct current.
Electrolyte	Conductive medium for exchanging charge internally between the positive and negative electrodes of a battery. In lead-acid and NiCd/NiMH batteries, the electrolyte is a liquid, which may be free, or absorbed in a separator, or combined with a neutral substance to create a gel.
Energy Density	In this book, we use *energy density* to mean a measure of how many watthours (Wh) per pound [or kilogram (kg)] are available from any given battery system. Batteries based on heavy metals, such as lead-acid, have low energy density, while those based on lighter elements, such as lithium, have high energy density. Also see *Power Density*.
End of Discharge	Minimum-allowed battery terminal voltage during a discharge. Also refers to the name of an alarm or supervisory control used to detect the end-of-discharge voltage.
End of Life	For a rechargeable battery, the end of its useful life is reached when it can no longer deliver at least 80% of its rated capacity during a discharge (after being fully charged). For a primary battery, end of life is reached when the item being powered no longer operates satisfactorily.
Equalize Voltage	A voltage that is higher than that required to bring a storage battery to full charge, but that is intended to equalize the level of charge in the series-connected cells of the battery. Also see *Float Voltage*.
Float Voltage	A voltage maintained on the terminals of a stationary storage battery designed to bring a discharged battery to full charge, and to maintain that charge during normal operation, while also supplying power to any connected equipment. Also see *Equalize Voltage*.
Flooded Cell	A battery cell with the active material (plates or grids) immersed in a container or jar of free electrolyte. Also called a wet cell. Most flooded cells have vent caps on the top of the battery jar to allow water to be added to maintain the electrolyte at the proper level, so the correct modern term is vented cell, to differentiate it from a VRLA, or

valve-regulated lead-acid cell, also called a sealed cell. "Sealed" is a bit of a misnomer, since VRLA cells have internal vents that are normally closed but can open to relieve excess internal pressure.

Hysteresis   A feature of an electronic circuit, such as an alarm, that prevents instability when the value being monitored is near the alarm point. For example, in a low dc voltage alarm, hysteresis prevents the alarm from toggling on and off due to load changes as the dc voltage decreases.

Hz   Hertz, or cycles per second, a measure of the frequency of an ac power source. In the US and most of the western hemisphere, that's 60 Hz. In most of the rest of the world, it's 50 Hz. Another reason we can't get along.

Lead Peroxide   An oxide of lead (chemical symbol PbO_2), usually mixed with other constituents, used to form the positive plate of a lead-acid battery. Widely referred to as lead dioxide.

Ohm   A basic concept in dc power systems. The ohm is defined using basic physical constants of time, length, mass and current. For our purposes, 1 ohm is the resistance, or impedance, of a conductor with a voltage drop of 1 volt, carrying a current of 1 ampere. Or, it's the resistance of a conductor that will pass 1 ampere when 1 volt is impressed on it. Resistance, in ohms, is measured with (of course) an ohmmeter.

In this text, we usually spell out the word ohm, but you may see the symbol Ω. Kilohms, kohms or kΩ, are thousands of ohms. Milliohms, thousandths of an ohm, are abbreviated as mohms, or preferably mΩ. Note the lower-case m. You may also encounter the terms Mohms or MΩ (upper case M), for megohms, or a million ohms.

Ohmic Isolation   In a circuit, this means that an ohmmeter would measure infinite resistance between two points in the circuit, such as the input and output of a power supply. In other words, there is no way for dc current to flow between those two points in the circuit.

Power Density   A measure of how much instantaneous power a battery of a given weight or volume can deliver. A battery may have high power density, yet only modest energy density. For example, an automotive SLI (starting, lighting, ignition) battery can deliver very high engine cranking power (easily over a thousand watts), but might have only a few hundred Wh of total energy.

Primary Battery
A battery type intended for a single use; that is, not rechargeable. The materials that comprise a primary battery are consumed during use and can't be reconstituted.

Ripple
Ripple is the ac component of voltage on the battery terminals during battery charging. It may also be measured on the dc output terminals of the charger if the battery is disconnected for maintenance.

Secondary Battery
A battery type intended to be used, or recharged, multiple times. The chemical reactions during discharge are reversible – the active materials can be reconstituted by proper charging.

Self-Discharge
The tendency of an electrochemical cell to lose useful capacity, even while not in service, usually through parasitic chemical reactions between the active materials, also called *local action*. In rechargeable batteries, most of the lost capacity can be recovered by recharging, if self-discharge hasn't progressed too far. If it has, some capacity loss is irreversible.

Specific Gravity
A measure of the concentration of the active chemical in a battery's electrolyte: potassium hydroxide (KOH) for nickel-cadmium batteries, and sulfuric acid (H_2SO_4) for lead-acid. Abbreviated sp. gr. Specific gravity is a relative number, with pure water being 1.0. If a fluid has a specific gravity of 1.25, then it's 25% heavier than water by volume.

Watthour
In dc systems, power is measured in watts. One watt is 1 ampere × 1 volt. If a battery delivers one watt for one hour, it's providing one watthour of energy. In mathematical terms, energy (watthours) is the time integral of power (watts). Abbreviated Wh.

ANNOTATED BIBLIOGRAPHY

The extensive list of standards for battery systems, including many from IEEE, referenced in Chapter 9, *Standards & Codes,* are not repeated below; only standards individually mentioned in other chapters are listed. Special-purpose standards, such as for nuclear generating stations, aren't included. You can browse the entire library of IEEE standards at http://standards.ieee.org/findstds/index.html.

There are many internet sites with information on primary and secondary batteries. Some are of questionable value, especially public Q & A forums. It's good to be skeptical. Wikipedia articles are usually good sources; other valuable sites are maintained by battery manufacturers. A selection of useful sites is listed below for the common battery chemistries. We don't list sites for lithium cells because the technology is evolving rapidly. To find information on lithium batteries, it's best to search for a particular type, such as "lithium ion secondary cell."

CHAPTER 1: BATTERY SYSTEMS — PRIMARY BATTERIES

"Eveready Carbon Zinc (Zn/MnO2) Application Manual." 2018. *Eveready Carbon Zinc Battery Handbook and Application Manual.* Energizer Brands, LLC. http://data.energizer.com/pdfs/carbonzinc_appman.pdf.

"Alkaline Manganese Dioxide Handbook and Application Manual." 2018. *Energizer Alkaline Handbook.* Energizer Brands, LLC. http://data.energizer.com/pdfs/alkaline_appman.pdf.

"Primary Cell." 2018. *Wikipedia.* Wikimedia Foundation, Inc. https://en.wikipedia.org/wiki/Primary_cell.
> *This article isn't of much use, although it does have a link to a useful table, "Comparison of Battery Types," which also offers links to many other cell types. Instead, try "Zinc-Carbon Battery," or "Alkaline Battery." These articles also provide links to some manufacturers' data.*

CHAPTER 1: BATTERY SYSTEMS — SECONDARY BATTERIES

2018. *Alcad.* Alcad. http://www.alcad.com.
>   Offers a wide range of documentation that may be downloaded, including
>   installation instructions.

2018. *Lead Acid Batteries.* Alcad. http://www.alcad.com/Products/Lead-Acid-batteries.
>   Alcad Lead-Acid Battery Operating Instructions are comprehensive and tailored
>   for each battery type.

2018. *EnerSys - PowerFull Solutions.* EnerSys. https://www.enersys.com/default.
aspx?langType=1033.
>   Allows downloading "short form" instructions and tabular performance data.

2017. *Document & Resources.* Exide Technologies. https://www.exide.com/en/
documents-resources.
>   Has a multitude of downloadable documents, many in languages other than English.
>   It's a huge site, so you have to have a clear idea of what product you need data for.

"Nickel Cadmium Batteries Application Manual." 2001. Eveready Battery Co. Inc.
http://data.energizer.com/pdfs/nickelcadmium_appman.pdf.

2018. *Cyclon Batteries.* EnerSys. https://www.enersys.com/Cyclon_Batteries.
aspx?langType=1033.
>   Discusses the Cyclon and Genesis battery brands. Offers downloads of a selection
>   guide, application guide, and MSDS.

Rusch, Weiland, Keith Vassallo, and Gary Hart. 2006. "Flooded (VLA), Sealed (VRLA),
Gel, AGM Type, Flat Plate, Tubular Plate: The When, Where, and Why. How Does the
End User Decide on the Best Solution?" http://ecreee.wikischolars.columbia.edu/file/
view/Rusch 2006 - Battery Type Overview.pdf.
>   Provides good quantitative comparisons of various types of lead-acid batteries.

"Stationary Battery Guide: Design, Application, and Maintenance: Revision of TR-100248."
2002. Electric Power Research Institute (EPRI). August 29. https://www.epri.com/#/pages/
product/1006757/?lang=en.

Williamson, Allan J, and John H Kim. 2015. "Comparison of Positive Grid Alloys for Standby Lead Acid Batteries." https://mc.services/wp-content/uploads/2015/12/Comparison-of-Positive-Grid-Alloys-for-Standby-Lead-Acid-Batteries.pdf.

> *Dives well into the nitty gritty.*

Szymborski, Joseph. 2018. "Self Discharge Mechanisms Occurring During the Open Circuit Storage of Valve Regulated Lead-Acid Batteries." https://www.researchgate.net/scientific-contributions/2057800982_Joseph_Szymborski.

> *There are two mechanisms which affect activated lead-acid batteries during extended periods of open circuit storage. The first is related to the corrosion of the positive plate grid, and the second is related to the self discharge reactions of the battery's active materials. Periodic charging during extended periods of storage will prevent any loss of lifetime or performance from grid corrosion and self discharge. The purpose of this paper is to review these two mechanisms and to discuss how GNB's ABSOLYTE® VRLA batteries withstand the conditions of open circuit storage.*

2018. *Basic to Advanced Battery Information from Battery University*. Cadex Electronics Inc. http://batteryuniversity.com.

> *Its tutorials evaluate the advantages and limitations of battery chemistries, advise on best battery choice and suggest ways to extend battery life. The information is compiled from specifications and independent test laboratories as well as crowdsourcing. Average readings are used when practical rather than citing research papers, as lab results are often not repeatable in real life.*

"IEEE Std. 1375-1998 - IEEE Guide for the Protection of Stationary Battery Systems." 1998. IEEE-SA - *The IEEE Standards Association - Home*. Institute of Electrical and Electronics Engineers. March 19. https://standards.ieee.org/findstds/standard/1375-1998.html.

"49 CFR 173.159a - Exceptions for Non-Spillable Batteries." 2011. *FDsys - Browse Code of Federal Regulations (Annual Edition)*. US Government Publishing Office. January. https://www.gpo.gov/fdsys/granule/CFR-2011-title49-vol2/CFR-2011-title49-vol2-sec173-159a/content-detail.html.

Darden, Bill. 2018. *Lead-Acid Battery Manufacturers and Brand Names List 2018*. June 13. http://jgdarden.com/batteryfaq/batbrand.htm#E.

O'Connor, Joe. 2017. "Battery Showdown: Lead-Acid vs. Lithium-Ion – Solar Micro Grid – Medium." *Solar Micro Grid.* Medium. January 23. https://medium.com/solar-microgrid/battery-showdown-lead-acid-vs-lithium-ion-1d37a1998287.
  Information regarding lithium ion manufacturing costs.

Climateer. 2018. "Lest We Forget, In April 2017 Platts Forecast Lithium Supply Would Outweigh Demand by 2018. Climateer Investing. February 19. http://climateerinvest.blogspot.com/2018/02/lest-we-forget-in-april-2017-platts.html.
  Large scale manufacturing cost of lithium cells was discussed in a news item from Platts in 2017. The Climateer article references the former article while providing an update.

Powers, Timothy. 2018. "The Most Advanced Hearables in the World Use Rechargeable Silver-Zinc Battery Technology." *ZPower Battery.* ZPower, LLC. June 13. https://zpowerbattery.com/advanced-hearables-world-use-rechargeable-silver-zinc-battery-technology/.
  Discusses hearing aid applications for silver-zinc cells.

CHAPTER 2: BATTERY CHARGERS

NEMA PE 5. 2004. "Utility-Type Battery Chargers." *NEMA - Setting Standards for Excellence.* National Electrical Manufacturers Association. October 25. https://www.nema.org/Standards/Pages/Utility-Type-Battery-Chargers.aspx.
  Float and equalize voltage ranges for lead-acid and nickel-cadmium batteries.

"Switched-Mode Power Supply." 2018. *Wikipedia.* Wikimedia Foundation. July 3. https://en.wikipedia.org/wiki/Switched-mode_power_supply.
  In-depth information on switched-mode power supplies.

CHAPTER 5: TEMPERATURE EFFECTS

"Molded Case Circuit Breakers Marking and Application Guide." 2016. UL, LLC. July. https://www.ul.com/wp-content/uploads/2014/09/CircuitBreaker_MG.pdf.
Identifies temperature requirements for circuit breakers used with battery chargers.

"UL - 489 Molded-Case Circuit Breakers, Molded-Case Switches, and Circuit-Breaker Enclosures | Standards Catalog." 2016. *Molded-Case Circuit Breakers, Molded-Case Switches, and Circuit-Breaker Enclosures.* UL, LLC. October 24. https://standardscatalog.ul.com/standards/en/standard_489_13.
The requirements of this standard cover molded-case circuit breakers, circuit breaker and ground-fault circuit-interrupters, fused circuit breakers, high-fault protectors, and high-fault modules.

"NFPA 70®." 2017. *NFPA.* National Fire Protection Association. https://www.nfpa.org/codes-and-standards/all-codes-and-standards/list-of-codes-and-standards/detail?code=70.
National Electrical Code for Circuit breaker ratings.

CHAPTER 6: ALARMS & OPTIONS - COMMUNICATIONS

1998. *TIA-485-A Electrical Characteristics of Generators and Receivers for Use in Balanced Digital Multipoint Systems.* https://global.ihs.com
Useful for biasing and termination resistor calculations and recommendations.

2006. *RS-422/485 Application Note.* B&B Electronics. June. http://www.atseries.net/PDFs/RS422 485AppNote.pdf.
Biasing and termination resistor calculations and recommendations.

"JD0062-00 Rev0C Ground Fault Detection Application Note." 2010. HindlePower Inc. February 3. https://www.hindlepowerinc.com/media/1423/jd0062-00.pdf.

"JD0036-00 Rev 10 Installing and Calibrating the Combined Alarm-Status Monitor (CASM)." 2009. HindlePower, Inc. October 30. https://www.hindlepowerinc.com/media/1328/jd0036-00.pdf.
Discusses CASM adjustments.

"Modbus Organization." 2018. *Modbus Specifications and Implementation Guides.* Modbus Organization, Inc. http://www.modbus.org/.

> *Details on the Modbus protocol.*

"Home - DNP.org - Distributed Network Protocol." 2018. *Pages — About Default.* DNP Users Group. https://www.dnp.org/default.aspx.

> *Details on DNP3 protocol.*

2017. *IEC 61850 Efficient Energy Automation with the IEC 61850 Standard.* Siemens AG. https://www.energy.siemens.com/br/en/energy-topics/standards/iec61850.htm.

> *Details on IEC 61850 protocol.*

NEMA. 2004. "Utility-Type Battery Chargers." *NEMA - Setting Standards for Excellence.* National Electrical Manufacturers Association. October 25. https://www.nema.org/Standards/Pages/Utility-Type-Battery-Chargers.aspx.

> *Differentiates between charger failure and zero current alarms and gives examples of failure causes.*

CHAPTER 7: APPLICATIONS

"946-2004 - IEEE Recommended Practice for the Design of DC Auxiliary Power Systems for Generating Systems - IEEE Standard." 2005. Wiley-IEEE Press. June 8. https://ieeexplore.ieee.org/document/1453057/.

Loehlein, Timothy A. 2006. "Calculating Generator Reactances." *Power Topic #6008 Technical Information from Cummins Power Generation.* Cummins Power Generation. December. https://power.cummins.com/sites/default/files/literature/technicalpapers/PT-6008-GeneratorReactances-en.pdf.

CHAPTER 8: WHEN BAD THINGS HAPPEN

"IEEE Std. 1375-1998 - IEEE Guide for the Protection of Stationary Battery Systems." 1998. *IEEE-SA - The IEEE Standards Association - Home.* Institute of Electrical and Electronics Engineers. March 19. https://standards.ieee.org/findstds/standard/1375-1998.html.

CHAPTER 9: STANDARDS & CODES

See an extensive list of standards and codes in the context of topics discussed in Chapter 9.

"White Paper #208 Understanding ANSI/IEEE C62.41 (Formerly IEEE 587)." 2018. *Downloads.* Powervar. Accessed July 9. https://powervar.com/downloads/.
> *For a deeper understanding of lightning and surge protection, download White Paper #208.*

APPENDIX A: FACTORY TESTING & ADJUSTMENTS

NEMA PE 5. 2004. "Utility-Type Battery Chargers." *NEMA - Setting Standards for Excellence.* National Electrical Manufacturers Association. October 25. https://www.nema.org/Standards/Pages/Utility-Type-Battery-Chargers.aspx.

APPENDIX B: USEFUL DATA

"IEEE Std. 1375-1998 - IEEE Guide for the Protection of Stationary Battery Systems." 1998. *IEEE-SA - The IEEE Standards Association - Home.* Institute of Electrical and Electronics Engineers. March 19. https://standards.ieee.org/findstds/standard/1375-1998.html

"485-2010 - IEEE Recommended Practice for Sizing Lead-Acid Batteries for Stationary Applications - IEEE Standard." IEEE Xplore Digital Library, Wiley-IEEE Press, 2010, ieeexplore.ieee.org/document/5751584/references..

INDEX

CREDITS

Content not specifically credited below either originates with the author, is the intellectual property of HindlePower, Inc. and used with their permission, or is information in the public domain.

1. Figure 1a, in Section 1.1.1.2, a cross-sectional diagram and an actual cross-section of a zinc-carbon cell, appears courtesy of Mcy jerry from https://en.wikipedia.org/wiki/Zinc-carbon_battery under CC BY 2.5 license.

2. Figure 1b, in Section 1.2.2.1, a drawing of the construction of a pasted-plate cell, appears courtesy of Tony Ortiz of HindlePower, Inc.

3. Figure 1f, in Section 1.4.7, a drawing of the cell-reversal process, appears courtesy of Tony Ortiz of HindlePower, Inc.

4. Figure 1g, in Section 1.5.2.8, a photograph of "Positive active mass fractioning," appears with permission of authors of "Electro-chemical Energy Storage," Petr Krivik and Petr Baca, which appears in Chapter 3 of "Energy Storage – Technologies and Applications" by Ahmed Faheem Zobaa, ISBN 978-953-51-0951-8, published January 23, 2013 under CC BY 3.0.

5. The two drawings at the start of Chapter 3 for the one-act play Ripple appear courtesy of Carlos A. Infante of HindlePower Inc.

6. Figure 30, in Section 3.5.2.3, on C-Message Weighting, appears courtesy of Lindosland through public domain licensing. It appeared in Wikipedia.org under "Psophometric Weighting."

7. Figure 6j, in Section 6.5.6.6, a schematic of an Ethernet network that isolates charger electronics using transformers, appears with permission of Pulse Electronics. It is based on a part available from the company at https://www.networking.pulseelectronics.com/.

8. Figure C1 in Appendix C, a schematic of a daisy-chained, four alarm relay, was designed by Matt Theriault of HindlePower, Inc.

ACKNOWLEDGMENTS

Bill Hindle, the founder and CEO of HindlePower, Inc., approached me in 2010 about writing a comprehensive manual on dc power conversion technology. When I pressed him for some definition of what he wanted the book to cover, his answer was always the same: "Write what you know." In other words, the content, organization, and target audience were open-ended. I hope that I've come close to satisfying his expectations.

I am grateful for Bill's encouragement and for his laissez-faire approach to managing the structure of this book. During the writing, his feedback was always positive. I will be pleased if you find the result to be up to HindlePower's exacting standards.

I cannot possibly thank everyone who has contributed, directly or indirectly, to my ability to complete this book, but some special acknowledgments are in order.

Bob Beck, the Chief Engineer at HindlePower, contributed the section on Communications, clarifying far better than I could the pesky details that must be mastered to create a successful network connection using Modbus or DNP3.

Several professionals reviewed the manuscript for grammar, syntax, and effusive overreaching – improving the consistency and readability of the text and keeping some of my enthusiasm carefully controlled.

Terry Guire, our editor at Brevity at Work LLC, worked tirelessly to get me to dot the i's and produce a completed manuscript on time. Terry was handed this project late in its evolution and has done a superlative job in creating a publishable product.

Finally, I thank all my associates at HindlePower for their friendship and support while I was learning from them. They are an outstanding crew, and it has been a pleasure to have worked with them for these past years. If this manual is useful, it is due in great measure to their help and influence. Of course, any omissions or errors in this manual are strictly my responsibility.

ABOUT THE AUTHOR

William K. Bennett's career in the field of electrical engineering has spanned more than 50 years. Since the 1960s, he has worked with organizations like the Navy, General Electric, Exide, Hitran, and others. At HindlePower, he led the development of electronic single-phase and three-phase microcontroller battery chargers and a very high-current dc power supply for plasma applications.

Bennett's latest book, *CHARGE!*, is a compendium of the knowledge he gained during his many years of hands-on experience with batteries and battery applications.

Now retired, Mr. Bennett lives in Yardley, Pennsylvania, where he enjoys crossword puzzles, listening to opera, and a late-evening dram of single malt with a little dark chocolate.